Field Guide for Managing Iron Sulfide (Black Powder) Within Pipelines or Processing Equipment

For Corrosion Control and Operations Personnel

Daniel E. Powell, Robert H. Winters, and Mark A. Mercer

Preface by Richard M. Baldwin

NACE INTERNATIONAL

NACE International
The Worldwide Corrosion Authority

Printed in the United States of America

ISBN: 978-1-57590-383-5

NACE International
The Worldwide Corrosion Authority
15835 Park Ten Place
Houston, TX 77084
nace.org

Dedication

This handbook is dedicated to all corrosion control and operations personnel, who are or have managed black powder within their pipeline systems or processing equipment.

Pyrophoric iron sulfide, whether within pipelines, vessels, or captured on in-line cartridge filters, may smolder or ignite when exposed to oxygen. This represents one of several potential hazards that corrosion control and operations personnel may encounter.

Table of Contents

List of Figures

List of Tables

About the Authors

Daniel E. Powell, P.Eng, FNACE

Before retiring, Daniel Powell focused his 40-year career in managing internal corrosion. He got his start in testing corrosion inhibitor packages for chemical cleaning of nuclear power steam generators, and later switched to providing internal corrosion control for the crude oil and natural gas production at ARCO Alaska, as well as Williams Companies. Inc., for their production, processing, and transmission pipelines. He has also provided consultancy services, again related to managing internal corrosion, and has authored over 30 publications. He is presently serving as an instructor for the NACE Internal Corrosion course.

Robert H. Winters, Sr. Pipeline Consultant

Robert H. Winters is a microbiologist and chemist, who has worked for El Paso Corporation, Tenneco, Inc., Tennessee Gas Pipeline Company, as well as Williams Companies, Inc. He also provided technical service consultancy for the crude oil and natural gas industry through Champion Technologies and is now providing those services through Clariant Corporation. The combination of operations experience, coupled with technical service consultancy, provides a diverse background for driving practical solutions to the challenges the industry faces. He has been a member of NACE for 40 years, has authored over 14 publications, and serves on numerous NACE committees for internal corrosion control.

Mark A. Mercer, Staff Technical Specialist

Mark Mercer is another 40-year veteran of the crude oil and natural gas production industry. He got his start in drilling operations, and subsequently transitioned to consultancy services for installing and maintaining cathodic protection, as well as coating, for both new and existing pipelines in Canada. He has also designed and implemented numerous chemical treatment programs for managing internal corrosion, as well as controlling scale formation. He has also led the efforts for establishing internal corrosion monitoring programs for several new fields.

Acknowledgments

A special thank you goes to each of the reviewers and their efforts to make this book become a useful tool for pipeline operations and corrosion control personnel. Besides verifying the technical accuracy of the contents, these individuals helped ensure each sentence was clear and understandable. We sincerely hope this book will become a useful tool for managing internal corrosion and in particular, black powder.

Richard M. Baldwin, Retired, Southwest Research Institute
Garry Matocha, Spectra Energy/Enbridge
Jack Smart III, Consultant
Virgil Wallace, Shell
Richard Eckert, DNVGL

Special thanks also goes to Daniel Purcell, Williams Companies, Inc., for his hard work supporting internal corrosion control and keeping black powder in check.

Additionally, special thanks also goes to Elaine Firestone, ERF Editorial Consulting, for taking the document from a technical draft, through NACE's publishing process.

Thanks to each of you for your time and dedication to the development of this book.

Preface

By Richard M. Baldwin, P.E.
Retired, Southwest Research Institute
August 2017

Richard M. Baldwin was one of the first investigators of black powder, which was found within oil and gas production pipelines and process equipment when problems were first recognized over 20 years ago. Using his mechanical engineering background, Richard consolidated bits and pieces of information to help craft some of the first publications on the subject. It helped create the recognition of the problem, from which the first initiatives for managing black powder were developed. This book is in part a follow-up to Richard's initiatives. Richard also provided a very thorough and comprehensive review of our text, which helped ensure technical accuracy and clarity. We are greatly appreciative of his time and commitment to sharing his knowledge and experiences in managing black powder.

–D.E.P.

In 1996, the Gas Machinery Research Council (GMRC) funded Southwest Research Institute (SwRI) to survey the gas industry to determine the variety of materials that have been found inside a gas pipeline. This interest was driven by the fact that many problems have occurred as a result of objects or materials, most of which move in a pipeline, propelled by the gas velocity. The result has been damage to reciprocating compressor rings, valves, and cylinder walls. In centrifugal compressors, the damage is in the form of erosion of blades. In either machine, the lubricating oil is sometimes absorbed or contaminated. Aside from machinery damage, some of the materials in the pipe have damaged—or caused inaccurate readings from—pressure sensors or measurement devices. The buildup of some materials is enough to actually reduce the gas flow, or require more compression work to keep it moving. Finally, some debris causes problems for the end-users where it clogs burners.

The survey found that there are many of the expected items such as weld slag and rods, dirt, salts, sand, water, hydrocarbon liquids, and even tools. However, the most unexpected and frequent component found, and which caused many problems, was a "black powder" that was sometimes dry like smoke or dust, and sometimes wet and tar-like. It was blamed for erosion of many moving parts in compressors, clogging of flow and pressure sensors, and building up to such quantities that it retarded flow. In general, it was a problem to maintenance personnel and a hidden cost to gas production.

We now know that black powder is a corrosion product, but at that time, it was not well known or understood. As a result of the survey's finding, the GMRC organized several forums attended by representatives of GMRC member gas companies. It was found that some people knew about the material, while others had very little knowledge of it—but no one was found who had enough understanding to prevent or solve the problem. As a result, the GMRC funded another two years of research to SwRI to find out about this material. What is it? Where does it come from? What can be done to prevent it? How should it be handled? How to get rid of it safely?

The result of this research was a document published by the GMRC for its members and available for free on its website: "Black Powder in the Gas Industry—Sources, Characteristics, and Treatment," Gas Machinery Research Council Report No. TA 97-4, May 1998, by Richard M. Baldwin. This report was followed by several magazine articles, follow-up papers, and presentations in pipeline-related publications. These are listed in the References [see references by Baldwin, and Baldwin et al.].

These documents revealed that the mysterious material, which that was black and had a number of components, was most often found to be predominantly one of several forms of iron sulfide. Iron sulfide is a corrosion product and the iron portion comes from the pipe wall itself. Unlike most problems that have a single root cause, this iron sulfide was found to have two sources. It could be formed either as a chemical reaction of the hydrogen sulfide in the gas with the iron in the pipe, or by certain microbes that assimilate sulfur from the gas, reside in a slime they build in water, and proceed to etch away the iron from the wall of the pipe, leaving this black residue and a pit in the pipe wall. This latter process is called "microbial influenced corrosion" or MIC. In both cases, the necessary ingredients are hydrogen sulfide and water. For either mechanism, the result is degradation of the pipe and a buildup of this corrosion product that causes the listed problems.

The second year of GMRC research also developed an insertion probe that could use a borescope to look for black powder inside a pressurized, operating pipeline. It further suggested possible means to insert a device to catch and extract particles for examination, measurement, and chemical testing.

Since the publications in 1998 and following years, knowledge of the problem has spread worldwide and others have entered the investigations. The corrosion engineers and pipe cleaning companies have advanced the state of the art in getting rid of the material. Operating companies have generated enough understanding that movements are now progressing to prevent the formation of black powder, facilitate its removal, and provide methods to identify its presence within a pipe.

This field guide has been prepared to provide corrosion control and operations personnel with the background for successfully managing black powder within their pipelines or processing equipment.

1 Introduction

Black powder is typically a mixture of iron oxides and iron sulfides, commonly found within pipelines transporting hydrocarbon products such as crude oil or natural gas, and may be found within the associated process equipment. The powder may be dry and in the form of a fine black powder or a slurry, when wetted by any produced water, lubricating oils from compressors, or perhaps carryover of glycols from dehydrators, which remove water during gas processing. The black powder may also be accompanied by formation fines from the hydrocarbon reservoir or scales, some of which may contain radioactive isotopes. Indeed, the compositions may vary. Although the first instinct is to collect and analyze a sample, the first priority needs to focus on how to *safely* collect a sample. Then, sample analysis and the establishment of a program to control black powder follows. This book is designed to assist the reader–corrosion control professionals and operations personnel–in recognizing and then managing and controlling black powder within their systems. Although black powder is the subject of this book, there will be significant discussions of iron sulfide, a component of black powder, simply because of the safety and health hazards posed by that chemical product.

Historically, black powder has been observed throughout the entire pipeline infrastructure, starting from upstream reservoirs and wells, to downstream transmission or sales lines, and finally the pipeline distribution networks that deliver the products to the customers. When samples are first collected from a new pipeline or gathering system, one of the first checks should be to determine if the material is radioactive due to naturally occurring radioactive materials (NORM). This would be checked using a Geiger counter, and if found to be radioactive, there would be different requirements for collecting and handling such radioactive materials. Likewise, if the powder sample was pyrophoric, meaning that it could burst into flames when allowed to dry and exposed to oxygen, then a different set of requirements would need to be implemented to ensure the safety of field personnel or laboratory personnel conducting further analyses.

Once the safety of personnel has been ensured through the use of personal protective equipment, then field tests can be conducted to provide a preliminary identification of the black powder. Is it magnetic? That is easy to confirm if the sample is dry; place a small sample inside a clear plastic sandwich bag and run a magnet across the surface. Magnetic materials will respond to the magnetic field. However, if it is wet, other factors, such as surface tension, may obscure any response. If the material is just a fine charcoal powder, there would be no response. Another quick test is to take a small sample about the size of a pencil eraser and wet it with a drop of hydrochloric acid. If it fizzes, the sample could be an iron oxide or an iron sulfide. However, if there is a smell of rotten eggs, then the sample contains at least some iron sulfide. Note that such tests should only be conducted in well-ventilated areas, and that the safest and best practice is to use the change in color of lead acetate paper to confirm the presence of hydrogen sulfide (H_2S), which is released when the drop of hydrochloric acid wets the iron sulfide powder.

Quick field tests can provide valuable guidance regarding the composition of samples when they are first collected. Additional field tests may include those to confirm the presence of water as well as to quantify the presence of any H_2S, or carbon dioxide (CO_2), which may be present within the gaseous phase.

This book was written to provide guidance for corrosion professionals and operations personnel in the proper handling of black powder from the point of initial sample collection and field analysis, through requesting the appropriate laboratory analysis. This book is also designed to provide guidance in establishing programs to manage the accumulation of black powder, such as through maintenance pigging or chemical treatments. The discussions would be applicable to gathering systems through transmission (sales) pipelines.

The obvious question is "why is it necessary to remove black powder from pipeline networks and process equipment?" Besides serving as an indicator of internal corrosion, the black powder that is formed or transported through a pipeline can plug any orifices and cause erroneous readings in gages or meters. Black powder has also been associated with erosion of metal depending on the velocities. Likewise, the presence of black powder may compromise quality control requirements. This book will have expanded discussions on each of these points.

1.1 When Black Powder was First Recognized as a Concern

A cursory review of papers associated with crude oil and natural gas operations yield virtually no references to black powder prior to approximately 1996, other than references to iron oxide and iron sulfide corrosion. However, concerns related to hydrogen sulfide and materials properties of steel pipelines date back to 1959 when engineers were first learning how to handle sour (H_2S containing) gas in western Canada. That is when many of those engineers organized as a NACE committee and developed the initial specifications for "Recommendations of Materials for Sour Service" (1B163), which was initially published in 1963. Ultimately, this document led to the now familiar NACE MR0175 standard,[1] which delineates "materials for use in H_2S containing environments in oil and gas production," first issued in 1975. That document is still regularly reviewed, updated, and approved as the standard to prevent sulfide stress cracking of metallic materials for oilfield equipment, such as steel pipelines that may be in sour service.

Although the industry has known for many decades the hazard associated with H_2S in crude oil or natural gas production, the focus had been the prevention of sulfide stress cracking of materials. It can only be speculated why there was not a secondary focus on the formation of black powder until the mid-1990s.

- Perhaps the newer techniques to increase production, such as water injection, which helped operators maintain or increase reservoir pressures, also introduced sulfate reducing bacteria (SRB) into those reservoirs. Those SRBs were then able to consume any sulfates present within the nascent reservoir or injected water and generate H_2S, where it may not have previously existed. That is how "sweet" reservoirs can become sour. The produced fluids/gasses would then contain some H_2S, which in turn would react with the iron within the steel piping, generating the iron sulfide characteristics of black powder.
- The accumulations of black powder within production systems may also have become more apparent as the flow rates were increased to meet customer demands. Perhaps those increased velocities were sufficient to transport the fine particulates.

Since the mid-1990s, there have been a number of technical papers published on the topic of black powder. These include publications by Richard M. Baldwin,[2-9] who focused on characterizing and controlling black powder; Abdelmounam M. Sherik et al.,[10-17] who focused on managing black powder within gas transmission pipelines; John S. Smart,[18-21] who focused on modeling particle size and the ability to transport black powder; and Steven M. Smith and Michael W. Joosten[22-23] and their review of technical literature on H_2S and CO_2 corrosion in 2006, which was updated in 2015 to reflect the most recent papers. Publications on microbiologically influenced corrosion, which relates to the microbiological generation of H_2S and the reaction with steel include those by B.J. Little and P. Wagner,[24] and D.H. Pope et al.[25-27]; Richard Eckert[28-30] provides in-depth discussion of all sampling and monitoring associated with managing internal corrosion.

Daniel Powell et al.[31] presented a paper that evaluated "black powder" generated in an on-shore gas processing facility. Although that "black powder" proved to be charcoal formed by overheating the incoming gas, the significance of that paper is that it illustrates the methodical process by which any powder sample could be analyzed, paving the way for remedial measures. The process will be discussed from the perspective of corrosion control professionals or operations personnel directing the field sample collection and selecting the laboratory analyses techniques for identifying the potential hazards, such that appropriate remedial measures can be implemented.

The diversity of papers also confirms that black powder is a common issue not only in the United States, but elsewhere in the world, including Saudi Arabia and China.[10, 32] It is a common concern worldwide. However, there are no previously published books, which primarily focus on and address the numerous topics related to black powder.

1.2 Goals for This Book

This book is attempting to provide a simple, straightforward overview of black powder, including what is black powder, how to analyze samples on site and in laboratories, and the establishment of programs to manage the accumulation of black powder through maintenance pigging and chemical treatments. By design, this book has been written to provide such knowledge for use by operations personnel and newer corrosion professionals who are in the field. The focus is practical to enable field personnel to quickly grasp the key concepts, while providing references, such that readers could further investigate the fundamentals. The following are the goals for this book:

- Provide a background regarding why black powder may be a concern
- Present safety and health concerns related to hydrogen sulfide gas
- Present safety and health concerns–pyrophoric iron sulfide and naturally occurring radioactive material (NORM)
- Provide potential indicators for the accumulation of black powder
- Define the sources of black powder, such that measures can be taken to prevent its formation
- Identify the numerous forms or structures of iron sulfide
- Discuss the collection of gas, liquid, and solids for analyses
- Discuss the types of analyses that should be requested for each type of sample and "why"–what those data analyses results will really tell us
- Provide an overview of maintenance/cleaning pigs
- Provide an overview of in-line inspections and identify limitations for magnetic flux leakage (MFL) or ultrasonic (UT) Pigs if the preexisting pits are "packed" with black powder
- Discuss internal corrosion monitoring, coupons and probes, and sulfide bridging, which can provide faulty readings
- Discuss black powder particle size and the velocities necessary to transport those particles through pipelines, including concerns related to abrasion or erosion
- Provide a strategy for identifying upstream sources of contaminants, e.g., H_2S
- Provide a strategy for establishing metrics for managing black powder

- Provide a case study from a pipeline cleaning using, tetrakis(hydroxymethyl) phosphonium sulfate (THPS), a chemical chelant and biocide
- Provide a case study of a pipeline cleaning using a combination oxidizer/ surfactant product
- Provide a case study of the cleaning of a natural gas liquids (NGL) tank using a combination enzyme, surfactant, oxidizer product

The many topics listed above all relate to the presence of H_2S, the formation and transportation of iron sulfide, the cleaning and removal of those corrosion products from pipelines and process equipment, and overall management of black powder. The discussion will start by describing what black powder is, and what the related health and safety concerns are.

Highlighting Most Significant Points

Grey boxes, such as this one, have been added to highlight the most significant points from those sections. This is intended to assist field personnel to locate key points.

2 What is Black Powder and What are the Health and Safety Concerns

Black powder is typically a combination of iron sulfide, iron oxide, and other contaminants, such as oil, water, salts, sand, and dirt, which are commonly found within pipelines or process equipment transporting hydrocarbons such as crude oil, natural gas, liquid petroleum gas (LPG) or condensates. The iron oxide and iron sulfide are corrosion products, which are typically associated with crude oil and natural gas production. Black powder can appear as a tar-like slurry in wet gas pipelines, or a very fine black powder in dry gas pipelines.

Besides being a contaminant, the fine particles of black powder can cause severe wear of compressor impellers or where they may impact pipe walls, such as sharp 90° bends. The black powder may also contain naturally occurring radioactive materials (NORM) scales, which could pose a health hazard to personnel who could come into contact with that material, such as when changing filters or removing pigs and debris from pig receivers following maintenance pigging.

This discussion will start with an overview of the numerous forms of iron sulfide, which is the most distinctive component of black powder. A discussion of health and safety concerns associated with H_2S and black powder follows. Note that H_2S, along with water, is essential for the formation of iron sulfides, and the presence of H_2S within pipelines not only presents a health and safety concern to personnel, but also a continuous generation of additional sulfides until the H_2S is totally consumed.

2.1 Numerous Forms of Iron Sulfide—A Component of Black Powder

There are many different forms of iron sulfide, which can form where the corrosion chemical reactions are occurring, notably within the aqueous phase, e.g., produced or condensed water where dissolved iron and dissolved sulfides mix. The subsections below describe the forms of iron sulfide as listed in a Pipeline Research Council International (PRCI) report.[8] The forms of iron sulfide are also presented in tabular form in Appendix A.

Consider the most common forms of the iron sulfide compounds, each of which can be distinguished from the others using x-ray diffraction techniques. It is important to know the particular form of iron sulfide that is found within individual systems. That enables an awareness of potential safety hazards, as well as facilitating the removal of those undesirable materials. Knowing the forms of iron sulfide present also allows for understanding the chemical conditions existing within the system, which were responsible for the formation of that sulfide.

The first three forms of iron sulfide discussed are pyrrhotite, troilite, and mackinawite. These are interesting minerals, because the ratio of iron to sulfur is less than 1.0 for pyrrhotite, equal to 1.0 for troilite, and greater than 1.0 for mackinawite. The vacancies of one atom or another, or the presence of other metal atoms within the crystalline structure, makes a difference in the stability and reactivity of these different forms of black powder.

2.1.1 Pyrrhotite ($Fe_{(1-x)}S$, where x = 0 to 0.2)

Pyrrhotite is a stable form of FeS, which has a bronze, bronze-red, or a bronze-red to dark brown coloration, and has been called "magnetic pyrite" because of the color and that it is weakly magnetic. It exists in a number of different hexagonal or monoclinic crystal symmetries—hexagonal structure when low in sulfur and closer to the 1:1 ratio for FeS, and monoclinic, meaning each axis of the crystalline structure has a different length when the structure is higher in the relative sulfur content. One common form of pyrrhotite has been designated as 4C (Fe_7S_8), where the 4C relates to the specific super lattice of the crystalline structure and the iron vacancies. The density of this compound varies, but is approximately 4.84 gm/cm^3.

Pyrrhotites have been synthesized in very short periods of time under laboratory studies through heating greigite and smythite in temperatures as low as 356 °F (180 °C). Pyrrhotite appears to be a more stable end product, compared to mackinawite, greigite, or smythite.

Whenever removing scales from a pipeline, whether through maintenance pigging or by replacing in-line filters, assume that the black powder is pyrophoric, and that the black powder is subject to spontaneous combustion when dry and exposed to oxygen. If the presence of pyrrhotite is either suspected or confirmed, it is best to keep the black powder removed from the pipelines moist or wetted, because the water serves as a heat sink to minimize the potential for spontaneous combustion.

2.1.2 Troilite (FeS), an "End" Member of the Pyrrhotite Group

Troilite is a nonmagnetic form of iron sulfide, which has the formula FeS and a density of 4.61 gm/cc. It is a hexagonal crystal, and, unlike pyrrhotite, does not have iron vacancies within the crystalline structure. As such, it is not magnetic. Troilite is rarely found on the Earth's crust or within pipelines. However, troilite is common within meteorites, as well as minerals analyzed from the moon and Mars, per space exploration programs.

2.1.3 Mackinawite ($Fe_{(1+x)}S$)

This is one of the first forms of iron sulfide that are formed in systems through the reaction of goethite (FeO(OH)) and hydrogen sulfide (H_2S). It is observed typically as a bronze or white-grey granular material, with the crystalline structure only apparent under a microscope. Note that nickel or cobalt may also be observed within the lattice. The composition of mackinawite is $Fe_{(1+x)}S$, where $x = 0.057$ to 0.064, with a density of 4.17 gm/cc. Mackinawite has an excess of iron, compared to the sulfur content, whereas troilite has an even ratio of iron to sulfur, and pyrrhotite has an excess of sulfur, compared to the iron content for that mineral. Mackinawite is most commonly observed in systems less than 284 °F (140 °C).

2.1.4 Pyrite (Cubic FeS_2)

Pyrite is a stable mineral, best known as "Fool's Gold," has a brass-yellow color and a metallic luster. It is chemically FeS_2, the same as marcasite, but has a cubic structure, whereas marcasite has an orthorhombic structure. Its density is 5.1 gm/cc. Pyrite forms when temperatures are greater than 302 °F (150 °C), and can be very protective. However, if the protective scale is broken through mechanical methods or chemicals, expect significant corrosion at the point of disruption.

Many years ago, naturally occurring pyrite used to be mined as a source for sulfur. However, sulfur is now more commonly obtained from the petrochemical industry.

2.1.5 Marcasite (Orthorhombic FeS_2)

This is chemically the same as pyrite (FeS_2), but has an orthorhombic crystalline structure and a density of 4.8 gm/cc. The typical crystal of marcasite has the distinctive appearance of a cock's comb, which will degrade slowly, emitting sulfur and causing corrosion of adjacent susceptible materials. The degradation appears to be triggered by the exposure to oxygen and is mildly exothermic. If marcasite is detected within any samples submitted for laboratory analysis, look for indicators of oxygen-related corrosion within that system. The major chemical product and service providers should be able to help in the selection of chemical products that can perform in the presence of oxygen.

2.1.6 Smythite (Rhombohederal $Fe_{(3+x)}S_4$)

Smythite, which has a greyish-yellow to reddish-brown appearance, is a strongly ferromagnetic material, which has the form of $Fe_{(3+x)}S_4$, where $x = 0$ to 0.3, and a rhombohederal structure (chemically Fe_3S_4). It is stable at high temperatures. Note that smythite can be a combination of iron and nickel, similar to pyrrhotite. As such, it is possible that nickel could be detected in X-Ray Fluorescence (XRF) or Energy Dispersive Spectroscopic (SRD) elemental analysis of a sample. Pipelines transporting produced fluids are typically fabricated from carbon steels, which do not contain nickel, making the presence of nickel less likely. However, if the presence of nickel is indeed confirmed, check the integrity of any stainless steel components within your production/processing systems. Smythite is slowly soluble in dilute hydrochloric or nitric acids.

Smythite has been synthesized through the sulfidation of siderite ($FeCO_3$). It can be partially transformed into pyrrhotite at temperatures of 356 °F (180 °C) in a water suspension within three hours.

2.1.7 Greigite (Cubic Fe_3S_4)

This forms from mackinawite and additional sulfide, like sodium sulfide. It typically forms at lower pH, between 2.3 and 6.5, and is generated more rapidly at lower pH. Warmer temperatures may also accelerate the chemical reaction forming greigite. It has not been possible to form greigite in direct reactions between iron and sulfur at any temperatures, suggesting that greigite may be a non-equilibrium phase between mackinawite and pyrrhotite.

2.1.8 Ferric Sulfide (Fe_2S_3)

Ferric sulfide is an unstable intermediate compound that transitions into a yellow-green powder ($Fe_2S_3 \rightarrow 2FeS + S$ (elemental sulfur) at ambient temperatures 68 °F (20 °C). If immersed in dilute hydrochloric acid, ferric sulfide reacts, forming iron chloride, distinctive hydrogen sulfide gas, and precipitates elemental sulfur:

$$Fe_2S_3 + 4HCl \rightarrow 2FeCl_2 + 2H_2S + S.$$

Ferric sulfide can be prepared in laboratories at cool temperatures, but will quickly decompose at ambient temperatures. It is not found in nature, and would not be expected in field samples.

There are many forms of iron sulfide compounds

- Pyrrhotite
- Troilite
- Mackinawite
- Pyrite
- Marcasite
- Smythite
- Greigite
- Ferric Sulfide

The reactions between the different forms are complex, and depend on numerous parameters, including temperature, pressure, and concentration of sulfides. Regardless of form, set up programs to monitor and remove those sulfides.

"The transformations from one crystalline form of iron sulfide to another are extremely complex and dependent on temperature, redox potential, relative concentration of iron and sulfide, as well as many other physical and chemical parameters. Some of these transformations can have a significant effect on corrosion by effecting changes in the continuity and protectiveness of protective films, by changing the potential of localized areas, or by electron transfer from anodic surfaces. The overall effects are not yet clearly understood by the corrosion community."[33]

2.2 Health and Safety Concerns Associated with H_2S and Black Powder

Section 2.2 presents an overview for safety and health concerns associated with hydrogen sulfide (H_2S) gas, pyrophoric iron sulfide, NORM, and abrasion or possible pipeline erosion associated with the presence of black powder. Hydrogen sulfide is addressed first, because it is the precursor for the formation of iron sulfides. Next, concerns related to the presence of pyrophoric iron sulfide will be addressed. Discussions of NORM, which may accompany deposition of iron sulfide and iron oxide, will follow. Concerns related to abrasion and possible erosion of the base metal of pipelines or process equipment, which may come from the transportation of black powder, are addressed in Section 2.2.5.

2.2.1 Health Hazards Posed by Presence of Hydrogen Sulfide Gas

Hydrogen sulfide (H_2S) is a colorless gas, which is corrosive and is very poisonous. It is most commonly characterized as having the odor of rotten eggs. H_2S can originate from the breakdown of organic material in anaerobic conditions, such as within swamps or sewers, and is typically found in crude oil and natural gas production around the world. It is slightly denser than air, and consequently uncontained releases of H_2S are more likely to accumulate at low points on the ground.

H_2S is soluble in water in low concentrations, up to 0.1 mole/liter at 64 °F (18 °C), and acts as a weak acid. H_2S also reacts with metal ions to form metal sulfides. One of the common tests to detect and confirm the presence of hydrogen sulfide is through the use of lead acetate paper. The H_2S reacts with the lead acetate to form lead sulfide, turning the lead acetate paper strips a dark brown to black color.

Per the Occupational Safety and Health Administration (OSHA), when exposed to low concentrations of H_2S, the exposed individual may experience irritation of the eyes, nose, throat, or respiratory system, although the effects may be delayed.[34-35] When exposed to moderate concentrations, there will be more pronounced or severe eye and respiratory effects, headaches, dizziness, nausea, coughing, vomiting, and difficulty in breathing. At high concentrations, which are concentrations greater than 100 ppm, an exposed individual may experience shock, convulsions, inability to breathe, coma, or even death. This can be very rapid, i.e., within a few breaths.

OSHA also termed those high concentrations as Immediately Dangerous to Life and Health (IDLH).[34-35] For that reason, personnel working in or near such concentrations of H_2S should be using the appropriate respiratory protection and personal protection equipment (PPE). Each company will have safety documents that define their requirements.

The U.S. Department of Labor also provides information on its website. This includes testing or monitoring the air for the presence of hydrogen sulfide, and training and educating individuals about hazards and controls. The Worker Exposure Limits[36] are presented in Table 1. Short term, acute symptoms and effects associated with exposure to different concentrations of hydrogen sulfide (H_2S) are presented in Table 2.[36]

Table 1 Worker Exposure Limits to H_2S[36]

NIOSH* Recommended Exposure Limits (REL) (10-min ceiling): 10 ppm
OSHA Permissible Exposure Limits: ▪ General Industry Ceiling Limit: 20 ppm (not to exceed at any time over an 8-hour shift) ▪ General Industry Peak Limit: 50 ppm (up to 10 minutes if no other exposure during shift) ▪ Construction 8-hour Limit: 10 ppm ▪ Shipyard 8-hour Limit: 10 ppm

* NIOSH: National Institute of Occupational Safety and Health (part of the CDC)

Table 2 Correlation of Concentration of H_2S Exposure to Symptoms/Effects[36]

Concentration (ppm)	Symptoms/Effects
0.00011–0.00033	Typical background concentrations
0.01–1.5	Odor threshold (when rotten egg smell is first noticeable to some). Odor becomes more offensive at 3–5 ppm. Above 30 ppm, odor described as sweet or sickeningly sweet.
2–5	Prolonged exposure may cause nausea, tearing of the eyes, headaches, or loss of sleep. Airway problems (bronchial constriction) in some asthma patients.
20	Possible fatigue, loss of appetite, headache, irritability, poor memory, dizziness.
50–100	Slight conjunctivitis ("gas eye") and respiratory tract irritation after 1 h. May cause digestive upset and loss of appetite.
100	Coughing, eye irritation, loss of smell after 2–15 min (olfactory fatigue). Altered breathing, drowsiness after 15–30 min. Throat irritation after 1 h. Gradual increase in severity of symptoms over several hours. Death may occur after 48 h.
100–150	Loss of smell (olfactory fatigue or paralysis).
200–300	Marked conjunctivitis and respiratory tract irritation after 1 h. Pulmonary edema may occur from prolonged exposure.
500–700	Staggering, collapse in 5 min. Serious damage to the eyes in 30 min. Death after 30–60 min.
700–1000	Rapid unconsciousness, "knockdown" or immediate collapse within 1 to 2 breaths, breathing stops, death within minutes.
1000–2000	Nearly instant death

Companies producing or processing crude oil and natural gas, as well as pipeline companies, should have safety programs directly related to the safe handling of the crude oil and natural gas products. These programs should have specific training related to the hazards posed by hydrogen sulfide or their reaction products, including black powder, which their workers may encounter. Company and site-specific rules and requirements should be followed.

2.2.2 Safety Concerns Posed by Presence of Pyrophoric Iron Sulfide

The safety and health concerns addressed in this chapter will bring to the reader's attention the pyrophoric nature of at least some forms of iron sulfide within black powder, and in particular that of pyrrhotite. "Pyrophoric" means that the material can spontaneously burst into flames when dry and exposed to oxygen.

Figure 1 illustrates a foam pig that has a burn mark, where a portion of the foam melted when a small, localized deposit of black powder evidently burst into flames and melted the foam after the maintenance cleaning pig had been removed from the pig receiver.

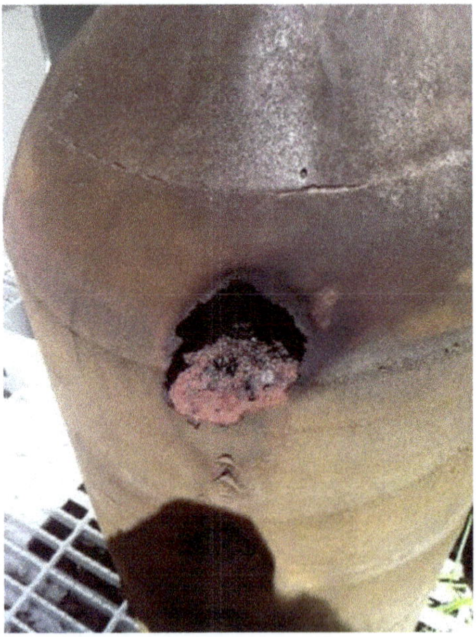

Figure 1 Maintenance Pig Damaged by Pyrophoric Iron Sulfide

However, note that even if maintenance cleaning pigs are removing black powder containing pyrophoric iron sulfide, they can be handled safely. Consider Figure 2, which illustrates a pig receiving facility that has been set up for safe handling of maintenance cleaning pigs from a pipeline, which had a history of pyrophoric iron sulfide depositions. When removed from the pig trap, the pyrophoric iron sulfide can quickly dry and burst into flames when oxygen is present. This photograph illustrates how one pipeline operator continuously wetted the interior of the pig receiver with water, including the pig when first received, and later, after the maintenance cleaning pig had been removed from the receiver. The lined pit at the pig receiver will capture both the water used to clean the pig receiver, and the water from the hose used to flush and clean the maintenance cleaning pig. The water from within the lined pit is then collected and taken to an appropriate hazardous waste site for proper disposal.

Figure 2 Pig Receiver and Containment for Handling Possible Pyrophoric FeS.

Water is directed into pig receiver to manage hazard from any pyrophoric iron sulfide.

Besides for helping to clean the pig and the receiver, the water serves as a heat sink for any pyrophoric black powder, which otherwise would burst into flame, after it had been removed from the pipeline and exposed to oxygen.

Next, consider filter cartridges used to capture particles, such as black powder, that would otherwise be transported through pipelines or process equipment handling nominally dry natural gas. The vessels housing the filter cartridges are typically found on the incoming line to a facility or just upstream of critical processing equipment. Operations personnel track the pressure differential across each of the in-line filters to determine when each of the filters needs to be replaced. Figure 3 illustrates a typical filter cartridge, which had been removed from service after the increase in differential pressure across the filter indicated the need for change-out.

Figure 3 Powder Accumulation Necessitated Replacement of this In-Line Filter

Used filter cartridges should be wetted as soon as they are removed from the process equipment, particularly if historical analyses of any solids samples collected from the process stream provides any previous indications of pyrophoric iron sulfide. Then, place the cartridges in air-tight containers to keep the filters and black powder moist while transporting those cartridges to an approved hazardous waste disposal site. Consult with environmental and hazardous waste specialists.

Figure 4 shows flames rising from a stack of filter cartridges, which were removed from service, set aside, and allowed to dry. The threat to health and safety from the spontaneous combustion of black powder, which contains pyrophoric iron sulfide, is apparent. Consider the possible consequences if such used filters were set aside within a facility, and there was even a small gas leak nearby. Alternatively, imagine the possible damage if those used filter cartridges were placed on the bed of a pickup truck. It has happened.

Figure 4 Black Powder Filters Fully Ignited[9]

2.2.3 Recognizing a Concern—Smoldering in the Field

Figure 5 illustrates smoke rising from the top of a slug catcher vessel at the inlet to a gas processing facility, after it had been shut down for maintenance and then opened to the atmosphere. The smoke is highlighted by the arrow in the photograph. Operations personnel recognized the increased differential pressure, which necessitated the filter change-out at the inlet to the facility. Pyrophoric iron sulfide was suspected, based on sample analyses from adjacent facilities, and consequently the operations personnel were prepared to address that potential safety hazard. Water was injected into the vessel to wet the filter cartridges and stop the smoldering, so that those cartridges could be replaced.

Figure 5 Smoke Vented through Open Top of a Slug Catcher Vessel

Figures 6 and 7 illustrate the smoldering of the filter cartridges, once they were removed from service, exposed to air, and had an opportunity to dry out. Note that water was sprayed into the dumpster to wet the filter cartridges and stop the smoldering until the cartridges could be properly disposed. Whenever there is a potential for the presence of pyrophoric iron sulfide, preparations should be made to wet the black powder until properly disposed. It is not acceptable to allow the filters to burn themselves out, because the product of that combustion is sulfur dioxide, a hazardous gas, which is referred to as "acid rain" when leached from the air by rain.

Figure 6 Dumpster with Smoldering Contents was Treated with Water

Figure 7 Filter Cartridges in Dumpster Still Smoldering

Keep Pyrophoric Iron Sulfide Wetted after Removal from System

- Keep cleaning pigs wet until they have been thoroughly cleaned to remove any pyrophoric iron sulfide
- Keep used filter cartridges wetted until properly disposed.

Refer to the company's environmental rules and procedures for the proper collection and disposal of hazardous waste. These rules will vary, depending on local, state/provincial, or national requirements, as well as the specific composition of the waste, including any radioactivity.

2.2.4 NORM in Black Powder or Debris Removed from Pipelines or Process Equipment

NORM can be a concern for some crude oil and natural gas production areas, whereas NORM is not a concern at other production areas. The concerns relate to whether the belowground production reservoirs contain NORM or not. This section raises awareness of the potential hazard and how to detect it, such that field personnel will be able to select the proper personal protection equipment (PPE) and work safely.

Radioactive material can be introduced into pipelines through upstream production, which carries the fine particulates of NORM into the pipeline gathering system and to, or through, the processing facilities. The deposits being removed from pipelines through maintenance pigging should be checked for radioactivity—particularly during the initial pigging runs, because any NORM associated with the deposits or corrosion products removed could pose a safety hazard to the individuals handling those materials. A recommended practice is to wear the appropriate PPE, including a respirator, when first opening a pig receiver or pig trap, following the initial maintenance pigging run. Then use a portable Geiger counter to check for any indications of radioactivity, which may have originated from any scales deposited into the pipeline. Respirators are recommended because of an increased health hazard if any radioactive particulates enter the respiratory system.

The priority for checks for NORM is on the first analysis of debris removed during the first maintenance pigging runs. That debris may include formation fines from upstream production, scales, and possible corrosion products. This check for NORM is to identify potential hazards that could affect the health of personnel handling the pigs and removing the debris.

If the presence of a potential hazard is confirmed, follow the company's Environmental, Health, and Safety (EHS) requirements, and any site-specific requirements for PPE to work safely.

Geiger counters are relatively simple devices for detecting the presence of ionizing radiation, but they cannot discriminate between different types of radiation, such as alpha or beta particles. Geiger counters also cannot provide data on the specific energy levels for the radiation, and they also are unable to accurately measure high radiation rates. As such, Geiger counters are designed for, and best used as, a screening tool to determine if there are potential hazards associated with NORM. If radioactivity is indicated, it is recommended that samples be collected for a more rigorous analysis, including determination of the specific type of radiation.

> Geiger counters should be used as screening tools to determine whether black powder or other debris removed from pipelines or process equipment contain naturally occurring radioactive materials (NORM), because NORM poses a threat to the health and safety of the personnel supporting those operations.

2.2.5 Abrasion: Concerns Related to Transportation of Black Powder Particles

Erosion or abrasion is a potential concern whenever black powder particulates are transported through pipelines. One article described the passage of the brittle particles of black powder as acting "like sandpaper, wearing the pipe wall."[37] Although black powder may be as large as 3.94 mil (100 μm) when formed, it will shear into fine submicron particles as a consequence of contacts with the pipe wall. A stream of such particulates can damage the impellor of a gas compressor in less than 24 hours.

Figure 8 shows the damage to a ceramic coated, gas cooled turbine blade, which was attributed to black powder particles that entered and struck the turbine blade. The forensic analysis reported the black powder evidently came from an upstream segment of the pipe, which had been pigged. The particles then traveled over 75 mi (121 km) downstream to strike the turbine blades.[21]

Figure 8 Abrasion to Ceramic Coated Turbine Blade, Attributed to Black Powder[21]

Once black powder enters a pipeline network, it will follow the direction of flow and can be distributed throughout the entire system. Although strides may be made in removing the bulk of that powder, expect at least a portion of that powder to remain within the pipeline network. Chapter 6 will discuss maintenance pigging and other programs to help remove black powder and other fine particles. Figure 9 illustrates similar efforts to remove black powder from a natural gas pipeline by blowing down a pipeline through a pig receiver at the end of that pipeline.[38] Note that this operation is not a recommended practice, but was included, because it illustrates inefficiencies in maintenance pigging for removing all particulates.

Figure 9 Blowing Down a Pipeline through Pig Receiver to Remove Black Powder[38]

Note: The blowing-down of a pipeline containing black powder is NOT a recommended practice.

This historical photograph represents the removal of a portion of the black powder remaining within a pipeline after completing maintenance pigging operation. The blow-down is a rapid depressurization, which initially increases velocities to sweep particulates out of the pipeline, and was intended to prevent the newly replaced filter cartridges from clogging up when operations resumed. It also illustrates inefficiencies that may be associated with some maintenance pigging operations.

2.2.6 Formation of Iron Sulfide May also Form Bridges or Compromise Isolation Gaskets

Section 9.3.4 discusses the monitoring of corrosion rates within pipeline systems or processing units. However, please note that if H_2S is present within systems that contain at least some moisture, iron sulfide can form. The resultant corrosion products may deposit on the elements of electronic probes used to measure corrosion rates. This phenomenon is called "bridging," because it can form "bridges," or an electronic short, between different portions of single element probes or between different elements of multi-element probes–that may result in invalid results. Likewise, if the electrically conductive iron sulfide deposition or corrosion products within a pipeline extend across an electrical isolation gasket, the functionality of that isolation gasket may be compromised and adversely affect the cathodic protection systems. Whenever there is a concern, local cathodic protection specialists should investigate each electrical isolation site.

2.3 Concentration of H_2S and Flow Rates to Estimate Potential Volume of Black Powder

H_2S is the essential ingredient for forming black powder, but requires the presence of water. This gas may come from the production reservoirs, or it can form from the microbiological activities of sulfate reducing bacteria (SRB) if they are resident within pipelines, whether from a recent hydrostatic pressure test of the pipeline, or perhaps accompanying produced water. H_2S is very corrosive, and will react with iron, such as pipeline steel, to form iron sulfide when at least some water (H_2O) is present. The specific operating conditions will determine which particular form or forms of iron sulfide are actually generated.

H_2S is the essential ingredient for forming black powder in pipelines.
H_2S will react with steel to form iron sulfide when water is present.

2.3.1 Estimating Quantity of Iron Sulfide That Can Form within a Pipeline

Consider a pipeline flowing 10 MMscfd (million standard cubic feet per day) of natural gas containing 4 ppm H_2S, which is a typical maximum specification for transmission-quality natural gas, and at least some moisture to serve as the electrolyte and support the formation of iron sulfide. Assume a 1 year period of time and that all of the hydrogen sulfide gas reacts with base metal on the interior of the pipeline. That small concentration of H_2S will generate approximately 3500 lb (1588 kg) of iron sulfide each year. Next, consider a flow rate of 100 MMscfd (2,830,000 m^3/d) for a two-year period of time. That same 4 ppm concentration of H_2S could potentially generate 35 tons (70,000 lb [31,752 kg]) of iron sulfide, assuming all the H_2S reacted with the steel in the pipeline. Such volumes of corrosion products would be problematic for any pipeline network.[39]

The balance is linear with time, concentration, and flow rate. There is no minimum concentration of H_2S, below which iron sulfide will not form–just acceptable rates. That is why it is essential to properly manage the black powder. Each pipeline operator needs to define their own limits, such that the volume of solids removed remains manageable.

Unless Managed, H_2S Generates Significant Volumes of Black Powder

Flow Rate		One Year at 4 ppm H_2S		Two Years at 4 ppm H_2S	
[MMscfd]	[m³/d]	[lb]	[kg]	[lb]	[kg]
10	283,000	3500	1588		
100	2,830,000			70,000 (35 tons)	31,750

Sidebar: How the Volume of Iron Sulfide Formed Can be Calculated

Assume the equation: $H_2S + Fe \rightarrow FeS + H_2$

Also assume, 1 mole of methane gas within the pipeline = 22.4 L at standard temperature and pressure (STP)

1 mole equals $6.02214199 \times 10^{23}$ molecules

1 ft³ = 28.3168 L

Hence 1 ft³ = 28.3168 L/(22.4 L/mole) = 1.264 mole

Flow Rate = 10 MMscfd

Flow Rate = 10,000,000 ft³/d × 1.264 mole/ft³ × 365 d/y = 4,613,600,000 mole/y

Given: H_2S is 4 ppm

Quantity of H_2S entering pipeline = (4/1,000,000) × (4,613,600,000 mole/y) = 18,454.4 mole H_2S

Per the above chemical equation, each mole of H_2S will form one mole of FeS, assuming complete reactions.

1 mole of FeS weighs 87.910 g (from Fe's 55.845 and S's 32.065 atomic weights)

Therefore, 18,454.4 mole H_2S forms 18,454.4 mole FeS
18,454.4 mole FeS × 87.910 g/mole = 1,622,322 g FeS

1 g = 0.00220462 lb

Hence,
1,622,322 g × (0.00220462 lb/g) = 3576 lb of FeS, or approximately 3500 lb, as above

2.4 Where Iron Sulfide and Black Powder Forms

Black powder can form throughout gathering and production facilities, from the wellhead through the production systems. It can also be generated in gas storage wells associated with the transmission systems. Integrity management programs should include steps for monitoring not only the concentration of H_2S and the moisture content, which are the precursors for forming black powder within carbon steel pipelines, but should also include processes and steps for managing its removal.

2.4.1 H_2S May Come from Reservoirs and Storage Caverns

Crude oil and natural gas from reservoirs are typically accompanied with produced water, which may contain some sulfates, H_2S, as well as SRBs and acid producing bacteria (APBs). The H_2S, coupled with any condensed or produced water, can result in the formation of black powder. As such, it is wise to monitor the concentration of H_2S, as well as the microbial population of SRBs and APBs coming from the reservoirs. Such monitoring provides the basis for whether upstream treatments may be needed.

The first step in processing the crude oil or natural gas is typically the flow into the slug catchers or first separators at those processing facilities. Those vessels are the first to separate any solids from the accompanying liquids and gasses. As such, they are subject to the accumulation of sand from the reservoirs, as well as corrosion products, including black powder. There should be options to clean out these vessels and remove accumulations of solids, including sand and black powder or other corrosion products.

If sales-quality hydrocarbon liquid or gas products are injected into storage wells or caverns upstream of the customers, be aware of the potential that the product retrieved from the storage wells, reservoirs, or caverns may pick up moisture, and H_2S may also move or drift above sales specifications. Reservoirs and caverns are typically cool and moist, and SRB within the reservoir are continuously generating hydrogen sulfide gas. The product may sometimes need to be treated to bring it back to sales specifications.

Also be aware that sulfur-containing compounds, such as mercaptans and thiols, are odorants, which are sometimes injected into pipelines, including some gathering systems. If individual pipelines are monitored using copper strips, which detect the presence of sulfides, the mercaptan and thiol odorants could provide positive indications, even if the pipeline does not contain any H_2S.

2.4.2 Process Upsets Occur—Hence, Continuously Monitor for Internal Corrosion

The processing facilities separate crude oil, produced water, and natural gas. However, such processing is not 100% efficient at all times. Upsets occur, and moisture or liquid water, may be inadvertently introduced into pipeline segments, which are normally dry. Such moisture is the electrolyte necessary for corrosion to occur. Hence, monitor and, as appropriate, treat each leg of the processing facilities. In addition, review results from inspections of any piping, vessels, or tanks for evidence of internal corrosion. Regular reviews of the results from monitoring programs should provide the data necessary for making any adjustments to the corrosion mitigation programs. Also, characterize the composition of any deposits, which were removed and analyzed.

2.4.3 Treat Waters Injected into Reservoir to Prevent Souring

The produced water removed during the separation processes is often reinjected into the production reservoir to maintain reservoir pressure and enhance the crude oil or natural gas recovery. However, if that water contains any SRB, such as from contamination of the produced water tanks at the separation facility, those organisms could reproduce and sour the reservoir because of their generation of hydrogen sulfide (H_2S). As will be discussed later, H_2S is the chemical that reacts with steel to form the black powder. Hence, the injection waters should be treated to control possible microbial contamination of the reservoirs, and the production and water injection pipelines should be pigged on a regular basis to remove depositions, corrosion products, or microbiological colonies attached to the pipe walls.

2.4.4 Process Gas to Specifications to Minimize Formation of Black Powder

The natural gas should be processed to sales specifications for export. Be sure that the composition of the gas, including the moisture content, and the concentration of carbon dioxide (CO_2), and H_2S, are continuously monitored.

Consider monitoring the total sulfur content. A measure for the total sulfur may also detect the presence of other forms of sulfur within the system, such as mercaptans or thiols, which are added in very low dosages as odorants. Take into consideration all sources for sulfur, as well as results from corrosion monitoring. Besides ensuring the gas meets the product specifications, such an approach also ensures that corrosion and the formation of black powder is unlikely.

2.4.5 Pig Sales or Export Pipelines Regularly to Remove Corrosion Products

Like gathering system pipelines, sales or export pipelines, which transport natural gas, should also be pigged at regular intervals to remove any corrosion products or black powder that may have accumulated within those particular pipelines. Compressor oil, or residuals of any glycols used in the dehydration process, might also be present. Although the gas product must meet specifications, be aware of the possibility for process upsets, which may allow for brief excursions from those specifications, such as for the moisture content. Also be aware of the possibility that the product could cool during transit, and consequently water could condense if the gas drops below the dew point. Such condensed water would be the electrolyte necessary for the corrosion mechanisms.

2.4.6 Change-out Filters and Conduct Maintenance Pigging to Manage Particulates

Sales or export pipelines transporting natural gas may also transport fine particulates of black powder, which have passed through the filters or bypassed them. These particulates can abrade the metal impellers of the compressors or pipe walls at pipeline bends. Consequently, the in-line filters should be replaced whenever the pressure differentials across the filters indicate the need for replacement. Be sure to verify the replacement filters are appropriately sized to capture any particles being transported through that pipeline. Additionally, the export pipelines should be pigged at regular intervals, such that the volume of material removed during each pigging run is kept at manageable levels.

The Bottom Line

- Black powder can be found at virtually any point throughout natural gas or crude oil gathering systems, processing facilities, as well as within sales or export pipelines.
- Use corrosion monitoring and sampling programs, to track the presence of black powder and quantify the volume over time.
- Establish metrics to assess the effectiveness of remedial measures.

2.5 Partial Pressure of H_2S and Sweet vs. Sour Service—NACE MR0175

The anticipated concentration of H_2S within a pipeline also should have been factored into the material selection for constructing the pipe itself. Pipelines built to transport hydrocarbon products, and which may be in sour service, should comply with NACE MR0175,[1] which addresses the concentration of H_2S within process gas, and the propensity for sulfide stress cracking of steels.

NACE MR0175, the industry standard for materials selections for "Sulfide Stress Cracking (SSC) Resistance Metallic Materials for Oilfield Equipment," provides a guideline for the concentrations of H_2S, above which MR0175 should be applied in material selections. SSC resistant materials should be employed when the partial pressure of H_2S in a wet (water as liquid) gas phase of a gas, gas condensate, or crude oil system equals or exceeds 0.0003 MPa absolute (0.05 psia).

NACE MR0175 Figures 1 and 2 illustrate sour conditions for gas and multiphase systems. Please refer to the NACE MR0175 document for guidance on materials selections. This document defines sweet or sour conditions, based on partial pressures of H_2S. Follow that document's guidelines for selecting materials to prevent sulfide stress cracking of metallic materials used in oilfield production equipment.

The concentration of H_2S within pipelines is significant for

- Rate of black powder formation
- Materials selection to prevent sulfide stress cracking (SSC)

Refer to NACE MR0175 for SSC Resistance Material Used for Sour Service

3 Additional Reasons Why Black Powder May be a Concern

This chapter highlights additional concerns related to the presence of black powder.

3.1 Transmission/Sales Lines—Ensuring Quality Control

Gas product specifications are established in the purchasing contracts, between companies. The specifications establish the acceptable and unacceptable quantities for each of the components, both major and minor, of the gas being transported. Typical specifications for natural gas being transported within transmission pipelines include specifications addressing moisture content, carbon dioxide (CO_2), the concentration of H_2S, which is usually specified in grains, and particulates. The specifications will vary, based on customers' requirements. However, the moisture content requirement for sales gas in the United States is typically expressed as 7 lb H_2O/million standard cubic feet (MMscf) (0.112 g/m^3). In Europe, the maximum permissible amount of water vapor within gas is based on the dew point, and is commonly set at −8 °C dew point, referenced to a standard pressure, such as 70 bar, where 1 bar equals the atmosphere pressure, or approximately 100 kPa. (Thus, a standard line pressure of 70 bar (1029 psi) would approximately equal 7000 kPa.) More rigorous requirements are typically required for offshore gas export pipelines—particularly those in deep water, such as in the Gulf of Mexico, where ambient temperatures can approach freezing conditions. Those specifications are typically 2–3 lb H_2O/MMscf ($3.2–4.8 \times 10^{-2}$ g/m^3), and are appropriate, considering the difficulty and expense to remediate any restrictions in flow, such as by the formation of hydrates, in the offshore deep water.

The CO_2 content specifications relate to the British thermal unit (BTU) content of the gas, and should also be monitored, because CO_2 will react with any condensed or free water to form carbonic acid, which is very corrosive. Note that a pipeline transporting pure carbon dioxide would not have internal corrosion problems if there is absolutely no water present, and there are a number of pipelines transporting pure carbon dioxide as part of enhanced oil recovery projects.

Hydrogen sulfide should also be monitored closely, because H_2S will react with the iron in carbon steel pipe or vessel walls to form iron sulfide, the primary component of black powder. The most common specification for the maximum allowable concentration of H_2S in sales or export pipelines is a quarter grain of H_2S per 100 ft^3 (where 100 ft^3 = 2.83 m^3) or approximately 4 ppm. This same concentration of H_2S was used in an example (Chapter 2) to estimate the potential volume of iron sulfide, which could form in a pipeline if moisture was present and all the H_2S reacted with bare base metal.

Finally, pipeline specifications for transmission-quality natural gas typically state the transmitted gas should be free of particles, whether in the form of solids or liquid droplets. The purpose of the in-line filters is to catch the fine particulates and protect the customer's equipment from faulty readings due to blocked meters or instrumentation.

Note that transmission-quality natural gas may get out of specification if temporarily stored in underground caverns or reservoirs. That is due to the moisture inherent in such caverns and the potential for generation of H_2S by any sulfate reducing bacteria, which are resident within those caverns. Evaluate the basis or root cause of each excursion on a case-by-case basis.

As an illustration of potential remedial measures, and using the example of temporary storage of transmission gas in a storage cavern or reservoir, consider an application of water-soluble H_2S scavengers to treat the hydrogen sulfide within the gas being retrieved from the underground storage caverns or reservoirs. Then pass that stored gas through a dehydration unit before being returned to the sales line.

3.2 Potential Corrosion, Wear, and Abrasion

Whenever particles or liquid droplets are being transported through pipelines, there is a potential for internal corrosion, as well as potential wear or abrasion. Base metals will typically develop a layer of corrosion products, such as an oxide, which slows down subsequent corrosion. However, solid particulates or high velocity liquid droplets can break established protective layers. When that occurs, expect the corrosion rate to initially be much higher when the fresh, bare metal first reacts with the local environment. Those corrosion rates will return to the previously established rates, once new protective layers are established.

Pipe walls at bends or tees, or processing equipment, such as pump impellors, may also be damaged because of being struck by numerous particles being transported through a system. Previously, Figure 8 illustrated the abrasion to a ceramic-coated turbine blade, which was attributed to black powder samples. Although not as distinctive, Figure 10 illustrates a gas compressor impellor that was removed from service after having been struck by a continuous stream of fine particulates of black powder following a pipeline pigging operation. Close-up, visual inspections are the key for confirming any abrasion or excessive wear.

Figure 10 Erosion Damage to a Pump Impellor Caused by Fine Black Powder Particles

One unique approach to monitoring the rate of metal abrasion would be to insert an inert metal coupon, such as a stainless steel coupon, into a system, and measure the weight loss over a period of time. Since the material is inert, any weight loss could be attributed to abrasion or wear from the particulates flowing through the process stream.[40–41]

3.3 Clogged Filters Reduce Efficiency and Increase Fuel Requirements

Black powder particles transported through pipeline systems can and will clog in-line filters. These filters will need to be changed out, once the difference in pressure between the inlet and outlet of the filters reach preselected values or pressure differentials, per the system's design. Such filter change-outs represent increased maintenance costs. However, those filters help protect downstream equipment, and also help ensure the natural gas being transported meets particulate specifications.

When in-line filters become clogged, there will also be some increased horsepower requirements to overcome the increased pressure differential across those filters. This can be tracked by virtue of the increased fuel requirements.

3.4 Iron Sulfide May or May Not Form Protective Scale

Some iron sulfides, like pyrite, appear to provide a protective scale. However, any disruption of that scale may result in increased localized corrosion where the scale was broken. Do not depend on protection from any pre-existing iron sulfide scale.

3.5 Iron Sulfide (Magnetic Particles) May Mask Pits in ILI Inspections

Pyrrhotite ($Fe_{(1-x)}S$, where $x = 0.0$ to 0.2), is a relatively stable form of iron sulfide and is magnetic, as are iron oxide (Fe_2O_3) particles. Both are common components of black powder. During maintenance pigging operations, fine particles of the black powder matrix are pushed down the pipeline and may be pushed into and fill pits on the interior surface of a pipeline, much like a paste filling holes. When magnetic particles fill the pits, the magnetic flux leakage (MFL) inspection tool might not be able to detect pitting, because magnetic fields can pass through the tightly packed magnetic matrix of black powder.

Prior to in-line inspections, efforts should have been made to clean the pipeline, using maintenance cleaning pigs, which have capability to scrape or "dig out" a portion of the debris within pre-existing pits. Section 7.3 addresses the use of a maintenance cleaning pig that have multiple spring loaded, rake-like fingers, which can "dig" debris from within pre-existing pits, making detection of such pits more likely. Also, consider maintenance pigs, which have attached magnets, for collecting magnetic debris from the interior of the pipelines.

Why Black Powder is a Concern

- Quality control concern for the transmission gas
- Potential wear and abrasion
- Increased ΔPressure across filter means more fuel needed for compressors
- Iron sulfide scales may or may not be protective
- May make pits more difficult to detect in ILI runs
- May form electrical shorts across the element or elements of corrosion monitoring probes
- Iron sulfide deposits on the interior of the pipeline may also form an electronic bridge across otherwise electrically isolated pipeline segments.

4 Potential Indicators of Black Powder in a System

This chapter describes indications that black powder may be accumulating in a system, and investigations should follow.

4.1 Increased Pressure Differential Across Filters

Production and processing facilities are designed to handle normal and upset conditions. In-line filters are commonly installed at the upstream entrance to a facility, or at critical points in the process. Operations personnel monitor the differential pressures across the filter, and when that differential pressure reaches a preset level, the filters are then replaced. Although expensive, the filters do protect the process equipment from damage due to particulates, or help ensure the pipeline specification for particulates is met. Figure 11 illustrates an in-line filter, which has been removed from service, and was replaced.

Section 2.2.2 provided a cautionary note regarding potential safety hazards posed if the filter is caked with pyrophoric iron sulfide particles, which can spontaneously combust if not kept wet.

Figure 11 Filter Removed from Service because of Increased Differential Pressure

4.2 More Frequent Change-out of Filters

The in-line filters used in the gas transportation industry are not simple, inexpensive filters, but instead are relatively expensive, and are sized, based on the design flow rates and the size of the particles to be removed. More frequent change-outs mean more expense, hence, it is beneficial to identify the source and minimize the volume of those particulates. Figure 9 in Chapter 2 depicts a pipeline operator's decision to "blow down" a particular pipeline, immediately following maintenance pigging runs. (The "blow down" is a rapid depressurization at one end of the pipeline to generate high velocities within the pipeline. As the gas escapes, the fine particulates are swept out of the pipeline.) The operator anticipated that if that particular pipeline had returned to normal production following the maintenance pigging, the fine particles remaining within the pipeline would have almost immediately clogged flow through in-line filters, necessitating an almost immediate replacement of those filters. Although *not* a recommended practice, the "blow down" also illustrated inefficiencies in their maintenance pigging operations.

4.3 Volume and Character of Material Removed During Pigging Runs

There are two potential indicators for changing conditions and possible formation of black powders.

First, look at the volume of material removed during any individual pipeline maintenance pigging run. Compare the volume removed during the most recent run to the volume removed during previous maintenance pigging runs. Any significant change, e.g., ± 10%, would warrant checks to determine if there were changes in operations or operating conditions, such as new production coming online.

> If the time intervals between the pigging runs are not the same, divide the volume of debris removed by the number of days between pig runs to determine a daily accumulation rate. Then compare accumulations over the same period of time.

Second, look for any changes in the characteristics or composition of the debris being removed from the pipeline. As a real-world example, one particular well, which was part of an offshore gathering system, contributed significant volumes of carbon dioxide (CO_2), as well as hydrogen sulfide (H_2S). When economics favored taking that particular well out of service, it was possible to significantly reduce the expenses for corrosion inhibitor chemicals, because the extra volumes of those chemicals were no longer necessary for controlling the corrosion. In addition, the presence of H_2S enhances CO_2 corrosion. Thus, the reduction and almost total elimination of H_2S would have the effect of reducing the corrosivity from the remaining CO_2.

4.4 Observed on Pipe Walls or Production Vessels When Open/Accessible

Whenever a pipeline or vessel is opened for making a pipeline connection, maintenance, or inspecting a process vessel, notify a corrosion professional so they can observe any internal corrosion and determine whether black powder is present.

> **Potential Indicators for Accumulation of Black Powder**
>
> - Increased pressure differential across filters
> - More frequent change-out of filters
> - Increasing volume of material removed during maintenance pigging
> - Direct observation whenever the interior of a pipe or production vessels can be accessed and observed.
>
> **Collect and analyze a sample to determine whether iron sulfide is present.**

5 Overview of Field and Laboratory Tests Used to Assess the Composition of Samples

The introduction provided a quick overview for testing to be conducted to characterize any samples of gas, liquids, or solids collected from pipelines or process equipment. The initial field tests (those conducted onsite), identify potential hazards onsite, which would necessitate additional Personal Protective Equipment (PPE) or other precautions. These would start with checks to determine if black powder or other samples contain radioactive material. The next field test would be to determine whether the material is magnetic, which relates to the risks associated with pyrophoric iron sulfides or the presence of iron oxides. Then, field tests can be conducted on a small portion of the sample to determine if sulfides are present. Again, these initial field tests are to identify the primary threats associated with the first sampling from new pipelines or process equipment, because that relates to personnel safety.

Field tests discussed in this chapter are to be conducted immediately after sample collection. This is necessary to (a) measure properties before they would change, such as pH of produced water before the outgassing of dissolved gasses shifts the pH, or (b) the inoculation of serial dilution bottles to quantify the population of any microorganisms, such as sulfate reducing bacteria (SRB) or acid producing bacteria (APB). Additionally, this chapter provides steps to preserve samples for analyses that are not time sensitive and would be conducted in a supporting laboratory.

This chapter will build on the above overview. The discussions in the subsections will be divided into gas, liquid, and solids samples. Each discussion will emphasize analyses appropriate to characterize the samples. Note that standard analytical techniques will be identified to make it easier to order the appropriate laboratory tests that quantify the composition of the samples, as well as to explain why complementary analytical techniques may be necessary. For example, when analyzing a dry black powder sample, X-Ray Fluorescence (XRF) would be used to determine the elemental composition, while X-Ray Diffraction (XRD) would be used to confirm the specific compounds, based on the crystalline structures.

This chapter will also include a case study, which was based on sampling from a new Colorado gas processing facility, and which illustrates the methodical approach to ordering sample analyses, and the need to recognize the limitations for some instruments. The latter is a reference to data collected using the energy dispersive spectrophotometry (EDS) capability of many scanning electron microscopes (SEM), which provides results similar to that from XRF. However, this particular SEM-EDS instrument did not have the capability to detect the presence of the element carbon. Sample analysis reports are based on percentages, such that the sum of all the components detected total 100%. Thus, the elemental composition and proportion reported for each component within a sample can be very different if one analytical instrument can detect more elements than a different analytical instrument used to examine the same sample. Significantly different results were obtained from analysis of the powder sample collected from this particular facility in Colorado—one instrument could detect the presence of carbon, whereas the other instrument could not detect the presence of carbon. That can lead to very different interpretations of the results and the appropriate remedial measures. *Thus, know the limits of the instrumentation, and ensure consistent results are obtained when reviewing all the data.*

The discussions in this chapter have been organized, based on practical considerations, including the most common field and laboratory analyses, as well as the relative cost and ease for conducting those analyses. Operations personnel want quick, accurate results to drive any decisions regarding possible remedial measures. Their focus is not on research projects. Hence, the most commonly used tools are noted first. Additional tools are noted, but with the understanding that they be employed for specialized analyses to diagnose unique problems.

Section 5.8 provides a graphical overview of the most commonly used field and laboratory analytical tools used to characterize samples collected for assessing black powder and identifying the potential causative mechanisms.

5.1 Collect Samples for Analysis of Components Associated with Gaseous Phase

Gas, liquid, and solid samples should be periodically collected to assess the various threats to the integrity for each system. The following subsections will address the collection and analysis of samples from the gaseous phase.

When assessing the potential for internal corrosion in wet or dry gas pipelines or process equipment, the first thing to do is to identify the potential for corrosive conditions. Hence, the initial analyses typically focus on measuring three parameters found in the gaseous phase:

(a) concentration of carbon dioxide (CO_2),
(b) parts per million of hydrogen sulfide (H_2S), and
(c) moisture content of the gas, which is typically expressed in pounds of water (H_2O) per million standard cubic feet (lb/MMscf), in North America. (Dew point temperatures, relative to a pipeline pressure, are commonly used in Europe as a measure for the moisture content.)

These measurements should be conducted in the field, as soon as possible, because the CO_2, H_2S, and moisture content of the gas can react and change in a short period of time.

5.1.1 Carbon Dioxide

If carbon dioxide is present in combination with condensed or free water, carbonic acid can form, which is extremely corrosive. However, if the pipeline is absolutely bone-dry, meaning that no water is present, internal corrosion will not occur, because there would be no electrolyte to complete the corrosion cell.

Sidebar: Partial Pressures of CO_2 Indicates Potential Corrosive Conditions

The pipeline industry uses a measure for the partial pressure of carbon dioxide in a pipeline as an indicator for the expected corrosiveness of that system. Partial pressure is essentially the portion of the pressure attributed to each component of the gas stream, based on the composition of the gas and pressure. Thus, if a gas contained 10% CO_2 and was at 200 psig (pounds per square inch gauge) (1380 kPa), the partial pressure of the CO_2 for that system would be 20 psig (138 kPa).

The partial pressures for CO_2 are used as an indicator for whether corrosive conditions would or would not be expected. Partial pressures below 7 psig (48 kPa) of CO_2 are generally indicative of noncorrosive conditions, while partial pressures above 30 psig (207 kPa) of CO_2 are generally indicative of probable corrosive conditions. Partial pressures between 7 and 30 psig (48 and 207 kPa) indicate that corrosive conditions may exist.

The general rules of thumb for corrosivity associated with different ranges of partial pressures of CO_2 are provided in Appendix A.

Measurement or knowledge of the CO_2 content of a gas stream is one step in assessing potential corrosivity in that pipeline network or processing equipment. This section notes one field technique for measuring CO_2 content; gas chromatography of samples collected in pressurized cylinders is another.

Field test kits are readily available to quantify the concentration of CO_2 in liquid samples, such as produced water. These are relevant, because CO_2 is highly soluble in water, and will quickly reach equilibrium between the liquid and gaseous phases. These liquid samples must be analyzed immediately after collection, since the CO_2 will quickly go out of solution and change the pH of the associated liquid.

The NACE Internal Corrosion–Basic course includes hands-on training for measuring the concentration of CO_2 dissolved in liquid samples, using specially prepared ampoules, shown in Figure 12, which use a sodium hydroxide titrant and phenolphthalene indicator to quantify dissolved carbon dioxide. These are based on ASTM D 513-82, "Total and Dissolved Carbon Dioxide in Water, Test Method E."[42] Most simply, this procedure for quantifying dissolved carbon dioxide calls for the filling of a small sample vial, adding a couple drops of an activator, stirring, and then drawing a small portion of the sample into the ampoule. The sample will immediately turn pink. Gently rock the ampoule to attain a uniform color, and then add additional sample until the liquid in the ampoule becomes colorless. Once colorless, the concentration is read by the level of the liquid, using the scale on the ampoule itself. This test methodology for measuring CO_2 dissolved in a liquid sample is a titration designed to be conducted on site, immediately after sample collection.

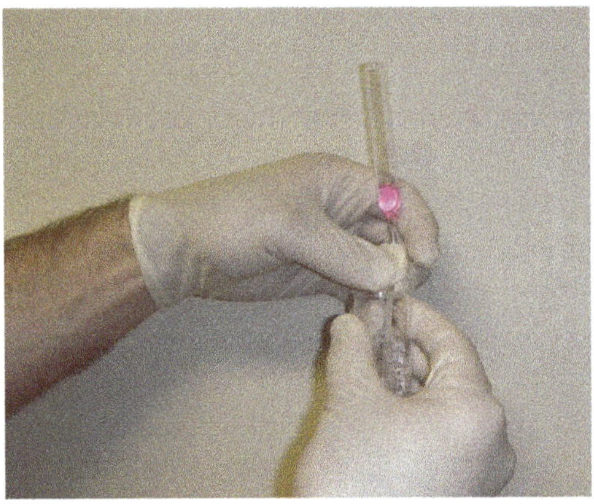

Figure 12 Phenolphthalene Indicator for Measuring CO_2 in a Liquid Sample[42]

Additionally, the composition of gas samples can be obtained through gas chromatography. Specifically, gas samples would be collected using pressurized sample cylinders, and then taking those samples to a laboratory for analysis. Besides for quantifying carbon dioxide, gas chromatography would also typically report any nitrogen, as well as methane, ethane, propane, butane, etc. If oxygen is reported, be suspicious of contamination during the sampling process, because oxygen would represent a very severe threat of system corrosion. Gas samples are collected at line pressure and temperatures, and they typically cool while being transported to the laboratories. Reheat or warm the test cylinders to operating temperatures before connecting to the gas chromatography units for analyses.

5.1.2 Hydrogen Sulfide

The concentrations of hydrogen sulfide (H_2S) in liquid or gas samples should be measured in the field using readily available colormetric test kits. The relative change in color equates to the concentration of the H_2S. Field analyses are essential for obtaining valid measures for the concentrations of H_2S, because H_2S is extremely reactive. If a sample were collected in the field and then taken to a laboratory for analysis, a portion of that H_2S is likely to have reacted, yielding lower concentrations of H_2S than would have been found in field sampling and analyses.

The NACE Internal Corrosion–Basic course includes a module for measuring the concentration of H_2S in water, using commercially available H_2S test kits. The first of these methods uses effervescent tablets to generate a fizz or bubbling of the test sample and lead acetate paper to quantify the H_2S. The use of lead acetate for testing the presence of hydrogen sulfide has been around since at least 1914, based on a Department of Commerce Technological Papers of the Bureau of Standards.[43] ASTM D 4084-07 (2012) addresses the test methodology, using lead acetate paper.[44]

The second method for measuring the quantity of dissolved H_2S in a liquid uses partially evacuated glass ampoules containing a solution of N,N-dimethyl-p-phenylenediamine and ferric chloride. Specific volumes of liquid samples are drawn into the ampoule due to the lower pressures of the partially evacuated ampoules. If those fluids drawn into the glass ampoule contain sulfides, the solution in those ampoules will turn blue with the formation of methylene blue. This is commonly referred to as the Methylene Blue Method, and is based on an EPA standard methodology.[45]

Stain tubes are used to measure the concentration of H_2S in the gaseous phase. The use of stain tubes will be discussed and illustrated in Section 5.1.3. Stain tubes are also available to measure the concentration of numerous other components in the gas, including the moisture content.

5.1.2.1 Effervescence and Lead Acetate Paper for Quantifying H_2S

The effervescent lead acetate test method for detecting and quantifying dissolved hydrogen sulfide starts by placing hydrogen sulfide test paper in bottle caps, and use tablets composed from sodium bicarbonate and anhydrous citric acid without aspirin (Alka-Seltzer® tablets) to generate the effervescence.[43] After the fizzing stops, match the color of the test paper in the interior of the bottle cap to the standard colors provided in the standard kits to determine the concentration of H_2S, as is illustrated in Figure 13.

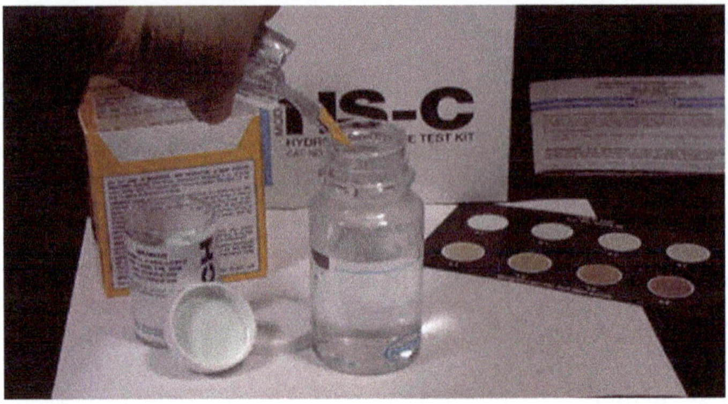

Figure 13 H_2S Measurement Using an Effervescent Lead Acetate Test Kit[43]

Test paper is placed in bottle cap, and later compared to color standards, shown on the right side of the figure.

5.1.2.2 Methylene Blue Method for Quantifying H_2S

The Methylene Blue Method test kits use ampoules, which draw a small, but fixed, volume of the liquid to be tested from a sample cup into the ampoule. After a set period of time, the color is compared to standards to quantify the concentration of H_2S. This is shown in Figure 14.

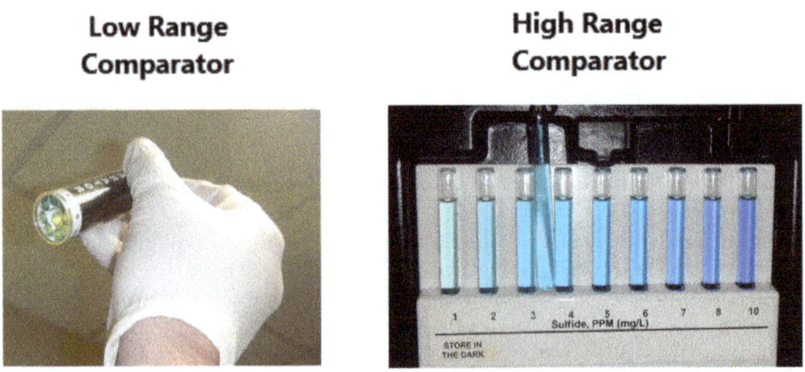

Figure 14 H_2S Measurement Using Methylene Blue Method Test Kit[45]
Compare color of sample ampoule to standards on the low or high range comparators

Field measurements, which show the presence of H_2S, provide strong evidence that iron sulfide black powder may be present in the pipeline or process equipment. For sales or export gas pipelines, the goal is to keep the concentration of H_2S at or below the sales specification, which is typically ¼ grain per 100 ft^3 (2.83 m^3), or approximately 4 ppm.

5.1.3 Measurement for Moisture Content or H_2S in Gas Using Stain Tubes

Field sampling and analysis should also include a measurement for the moisture or water content, as well as a measure for the concentration of H_2S in the gaseous phase. These analyses can be conducted in the field using stain tubes. Stain tubes are small glass tubes filled with a specially coated resin, which will react to the presence of the gaseous phase components of interest, such as water vapor or H_2S. (Numerous different types of stain tubes are commercially available, depending on the component of interest and the expected concentrations.)

The stain tubes have very fine points at each end of the tubes, and the ends are snapped off just before taking the measurement. One end is inserted into a hand pump, while the opposite end is inserted into the test stream. Thus, a controlled volume of the gas is drawn through the stain tube with each squeeze of the hand pump. The concentration of the component of interest, e.g., moisture content or concentration of H_2S, is determined by the length of the discoloration or stain, as measured on the stain tube, when a controlled volume of the gas sample is drawn through that stain tube. Note that the manufacturers place marks on the exterior glass of the stain tubes, and those marks are used to identify the concentrations for the component being tested.

Figure 15 illustrates how a handheld pump, coupled with a stain tube, can be used for field measurements of specific components within a gas stream. Note that the technique is nowhere near as precise as results, which could be obtained from analysis of samples collected in gas cylinders and taken to a laboratory for analysis. Nevertheless, results from stain tube measurements can be used to provide a quick estimate for the concentration of any particular component of interest. The figure illustrates a plastic bottle, which has been inverted and filled with the sample gas through the bottle cap creating a localized environment that approximates the conditions within the pipeline. The plastic bottle has holes to allow a stain tube to be inserted into the sample gas. The holes are sized to prevent over pressurization. Figure 16 illustrates a change in the color of a stain tube, which provides a measure for the concentration of the chemical component of interest, here being H_2S.

Figure 15 Hand Pump and Stain Tube for Measuring Components in Gas
The plastic bottle above the valve provides a localized environment for sampling.

Figure 16 Measurement of Component in Gas, Using Hand Pump and Stain Tube
Commercial suppliers offer a variety of stain tubes, each for measuring the concentration of different components of the gas stream, such as H_2S or moisture content.
Photograph courtesy of Clariant Oil Services

Moisture measurements using stain tubes are essentially identical to the techniques for measuring the concentration of H_2S, with the exception that different stain tubes are selected. Commercial manufactures offer stain tubes capable of measuring the component of interest and the expected concentrations.

Measuring the moisture content of natural gas pipelines is important, because the moisture content, coupled with knowledge of the line temperatures and pressures, can be used to estimate the possible volume of water, which can condense as temperatures drop. Condensed water is the electrolyte necessary to support corrosion mechanisms.

The moisture contents for air in Houston, Texas can be in the range of 350–400 lb H_2O/MMscf (5.6–6.4 g/m^3). Gas gathering pipelines may exceed 50 lb/MMscf (0.80 g/m^3), while gas sales or export pipelines typically keep moisture content below 7 lb/MMscf (0.112 g/m^3). Water condensation is possible, depending on the composition of the gas stream, temperature, and pressure. Iron sulfides can form if the gas stream contains any H_2S and there is any free or condensed water.

Knowledge of the moisture content of the gas in a pipeline is essential to controlling potential internal corrosion. This is typically accomplished by either preventing condensation, or through the application of chemicals, such as corrosion inhibitors. Consider modeling the moisture content for gas flow through individual pipelines or pipeline networks, and take into account changes in temperature and pressure.

There are no internationally accepted standards for the moisture content of natural gas sales/transmission pipelines. However, the most commonly used pipeline specification for considering a pipeline to be "dry," is when the moisture content is kept at or below 7 lb H_2O/MMscf (0.112 g/m^3).

Several pipeline operators in the deep water of the Gulf of Mexico have set a more restrictive pipeline specification of 2.5 lb H_2O/MMscf (0.04 g/m^3). Their concern is that the temperatures at the deepest points on the bottom of the Gulf of Mexico are approximately 32 °F (0 °C), and they want to prevent any condensation of moisture during transit to the onshore processing facilities. Besides for controlling corrosion, there is also the concern related to the formation of methane hydrates from chilled methane and water. It is very expensive to remediate (remove) any hydrates formed offshore, particularly those in deep water on the bottom of the Gulf of Mexico, where there are limited access points for injecting chemicals to prevent hydrate formation. Also keep in mind that minimizing moisture content reduces the potential for black powder formation.

As noted above, the concentration of H_2S in a gas stream can be measured using stain tubes. The most common specification for sales or transmission-quality natural gas is ¼ grain per 100 ft^3, or approximately 4 ppm. Stain tubes are an excellent tool for quickly monitoring and assessing this particular threat.

Gas Analyses Provide Valuable Data to Identify Potential Corrosive Conditions

- CO_2 (corrosive in water; partial pressures of CO_2 correlate to potential corrosivity)
- H_2S (enhances corrosion mechanisms; forms black powder)
- Moisture Content (electrolyte for corrosion; formation of hydrates)

Samples react and degrade quickly. Hence, collect and analyze immediately!

These three measures can provide a good understanding for potential corrosion or the formation of iron sulfide, the distinctive component of black powder.

5.2 Collect Samples for Water Analysis

Water collected from sample points along pipelines, from pig traps, or from within processing facilities can provide much valuable information regarding the ongoing corrosion mechanisms. The most common analyses would be (a) elemental analysis, using atomic absorption instruments to quantify elemental compositions, such as iron (commonly called "iron counts"), barium, etc., which are typically associated with the formation of scales; (b) wet chemistry or ion chromatography to quantify various cations, such as chlorides, sulfates, etc.; and (c) serial dilution or other techniques to quantify microbiological populations. The remainder of this section will present a discussion on the elemental analyses and wet chemistry, and the purpose or significance for collecting the data. Section 5.3 will present an overview for the numerous techniques used to detect and quantify microbiological populations.

5.2.1 Elemental Analyses Using Water Samples and Atomic Absorption Technique

Elemental analyses of liquid samples are invaluable, because the results will provide insight regarding underlying corrosion mechanisms. Iron, chromium, and nickel are of interest since elevated concentrations of these cations are typically indicative of corrosion of carbon or stainless steel. If copper is detected, that suggests investigating corrosion mechanisms associated with any components containing copper. Calcium or magnesium may suggest scale formations. The presence of phosphorous suggests the presence of corrosion inhibitors containing phosphate or perhaps scale inhibitors containing phosphonates, and so on. The elemental analysis of liquid samples may provide insight into corrosion mechanisms, and hence should be one of the first tests undertaken when analyzing the first samples from pipelines or process equipment—especially those reporting black powder. Atomic absorption is one of the first, generally recommended tests, because it is cost-effective, and laboratories supporting pipeline operations typically have such instrumentation available.

Atomic absorption (AA) can quantify the concentration of virtually any element in a liquid sample. However, the instrument has to be programmed or set up to look for specific elements of interest. Otherwise, other elements would not be detected and reported. AA analyses is commonly used for quantifying the concentration of iron, the principal component of carbon steel pipelines, or elements such as barium, which is associated with a particularly tenacious form of scale found within the petrochemical industry.

NACE SP0192 2012, Monitoring Corrosion in Oil and Gas Production with Iron Counts, would utilize data from liquid samples, and generated through the use of AA instrumentation.[46] Note, however, that the AA technique is to quantify the concentration of iron or other elements in aqueous solutions. Results from the AA instrument would be expected to complement those from other techniques, such as from XRD studies of solid samples collected from the same location, and which will be discussed shortly.

The AA technique is highlighted first, because many laboratories have these instruments. However, if there are very low concentrations of the element of interest, more sophisticated and expensive techniques should be considered. Gas Chromatography–Mass Spectrometry (GC-MS), can provide results in parts per billion. Such precision can be very useful when diagnosing and confirming corrosion mechanisms. Consider, for example, the diagnosis of manganese (Mn) in water samples. Pipeline steel has a characteristic ratio of iron (Fe) to Mn. If that steel corrodes, water samples collected from the pipe would be expected to contain both Fe and Mn in approximately the same ratio as in the steel specifications. However, nascent formation water from the crude oil or natural gas reservoir commonly contains some Fe. Thus, the presence of Mn can be used as an indicator for active pipeline corrosion. (Manganese is rarely, if ever, detected in produced water from reservoirs, and is a good indicator for corrosion of pipeline steel. The concentrations of manganese to iron should match the relative proportions in the base material if corrections are made for the concentration of any iron from the nascent reservoir water.)

There may be several days to weeks from the time of sampling to the time of analysis, and some of the ions may precipitate, or come out of solution, prior to analysis. Hence, a standard practice to keep any components of a test sample from precipitating is to add a small volume of hydrochloric or nitric acid, to keep those ions in solution. Consider placing 5 mL of (1.0 normal) hydrochloric acid or nitric acid into a 100 mL sample bottle, and then filling the bottle to the 100 mL mark with the fluid sample. Such "acidification" of fluid samples is to prevent individual cations from precipitating before laboratory analyses are completed.

5.2.2 Wet Chemistry to Quantify Specific Cations

There are also numerous wet chemistry techniques that could be used to detect and quantify the concentration of ions present in a liquid sample. These would include straightforward chemical titrations to various chromatography techniques. Crude oil and natural gas producers and pipeline operators rarely have laboratories with sophisticated laboratory equipment. However, most of the major chemical product and service providers have field offices, which can conduct routine, simple analyses. As warranted, individual samples may be forwarded to regional or headquarter laboratories for more sophisticated analyses by specialists.

5.2.3 Many Other Analytical Instruments—Check Capabilities of Supporting Labs

There are a number of instruments that can detect and quantify the presence of various anions, such as chlorides, sulfates, sulfides, etc. Such techniques include ion chromatography (IC) or ion exclusion chromatography (ICE). New techniques are continuously being developed. Consult local, supporting laboratories to identify their analytical capabilities.

5.3 Detecting and Quantifying Microbiological Populations of SRBs

The two most commonly used approaches to detecting and quantifying microbiological populations of sulfate reducing bacteria (SRBs), which generate H_2S, are (a) the serial dilution, and (b) the adenosine 5'-phosphosulfate (APS) reductase field tests. These two approaches will be discussed, because they are the primary tools associated with evaluating the populations of SRBs. A few additional techniques will also be presented. However, those additional techniques are used to quantify general microbiological populations, rather than those specifically designed to quantify SRBs, which generate H_2S and directly affect the formation of black powder. The comments in Section 5.3.3 will identify the test methods that quantify all bacteria, as opposed to those exclusive for SRBs.

Best Technique for Assessing SRB Populations. Use APS reductase and or serial dilution test to quantify SRB populations

- Serial dilution is most sensitive, but takes 28 days to yield results
- APS reductase is less sensitive, and cannot distinguish between live SRBs and those freshly killed, such as immediately following a biocide treatment. However, results are available in ~1 hour. Best option if time critical analyses are needed.

5.3.1 Serial Dilution

Serial dilution studies are used extensively throughout the oil and gas production industry to quantify the presence of SRBs or APBs. Typically, serial dilution tests are conducted simultaneously for determining the populations of both SRBs and APBs. The bottles used for assessing populations of SRBs contain a sterile nutrient media for SRBs, while the bottles used to assess populations of APBs contain a sterile nutrient media for APBs. The serial dilution studies start with the collection of a sample and separately injecting 1 mL from that sample into the first bottle for each series. Specifically, 1 mL of the sample solution is placed in the first bottle of each series of nutrients, each of which initially contain 9 mL. Thus, when the 1 mL sample is added to the first serial dilution bottle, a 1:10 dilution is achieved. Next, 1 mL is taken from the first bottle in the series (which contained 10 mL after the sample was added to that bottle), and placed into the second bottle of the series, which previously contained 9 mL of sterile solution, achieving a second 1:10 dilution from the first bottle, or the desired 1:100 dilution of the original 1 ml sample. Then, 1 mL is taken from the 10 mL in the second bottle and injected into the third bottle of the series. Thus, a 1:10 dilution of a 1:10 dilution of a 1:10 dilution of the original sample achieves the desired 1:1,000 dilution, and so on. Positive responses, as indicated by color changes within the bottles after the established incubation periods, are used to quantify the microbiological population within the original sample.

The bottles are then placed in a dark incubator kept at approximately the same temperature as where the samples were collected. The bottles are then checked at a specific exposure time to determine whether the color of the nutrient bottles has changed or not. The number of bottles changing color in each series correlates to the specific microbiological populations for the SRBs or APBs and the accumulation of their waste products, which changes the pH of the solution and the apparent color. Please refer to NACE TM0194-2014 (or most current version) for a detailed description of the technique.[47] The techniques are also part of the hands-on training in the NACE Internal Corrosion–Basic course.

Figure 17 illustrates the collection of a sample drawn from a sample port on a pipeline, while Figure 18 illustrates a sample being collected from a pig receiver.

Figure 17 Collecting a Sample from a Sample Port
Photograph courtesy of Clariant Oil Services

Figure 18 Obtaining a Sample from Receiver

Figures 19 and 20 illustrate the collection of 1 mL from the sample, and injecting it into the first bottle. Subsequently, 1 mL will be obtained from the first bottle and injected into the second bottle, with each transfer representing a 1:10 dilution.

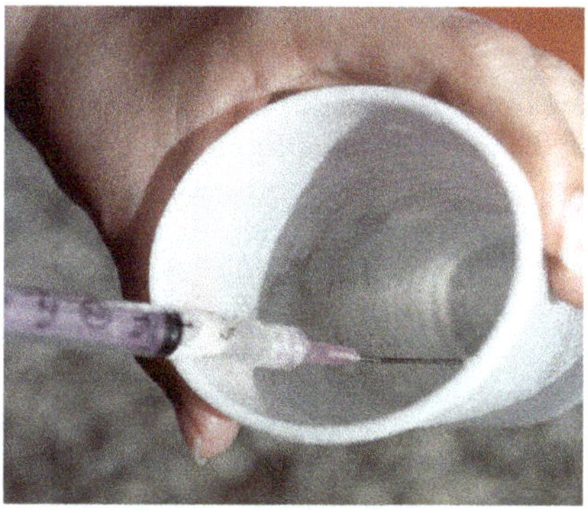

Figure 19 Collect 1 mL from the Original Sample for Serial Dilution Studies

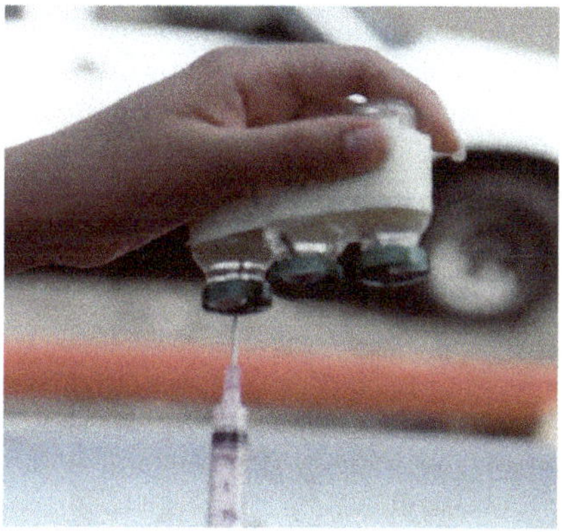

Figure 20 Serial Dilution Illustrated
Transfer 1 mL from first to second bottle; then 1 mL from second to third bottle.

5.3.1.1 Precautionary Notes for Generating Good, Valid Sampling Results

Once a sample has been collected, the serial dilution of that sample should be undertaken as soon as possible to accurately quantify microbiological populations of the SRBs. Otherwise, the environment in which those microorganisms lived could change, yielding inaccurate results. Direct sunlight could also kill microorganisms, resulting in faulty results. Hence, complete the serial dilutions as quickly as possible after collecting the original sample. Then place those bottles into an insulated chest to maintain the same temperatures as when collected, and keep them out of heat or cool conditions and direct ultraviolet sunlight.

5.3.1.2 Reading the Vials

Next, consider how the sample bottles or vials are read. As stated above, solutions in each of the vials have been especially prepared for the microorganisms that are to be quantified. The vials should be kept in a dark incubator, which is maintained approximately 9 °F (5 °C) of the typical operating temperatures for the system at the sample collection point. The microbiological population in each of the nutrient vials will increase, but so does the volume of their waste product.

There are also specific time intervals for the number of days from the sample inject into each of the bottles, to the time at which the populations are quantified. The total incubation time for the bottles used to quantify SRB populations is 28 days, whereas the total incubation time for bottles used to quantify the population of acid producing bacteria is 14 days.

If the color of the solution in any of the bottles used for testing SRBs changes from "clear" to "black," the vial is reported as positive for the presence of SRBs. EXCEPTION: If the color change is almost immediate, e.g., minutes to less than 2 hours, the test would be considered invalid, due to the apparent concentration of dissolved hydrogen sulfide (H_2S) within the sample itself. Consider other techniques to quantify the presence of SRBs.

Vials that have not changed color, are considered as "negative" only after completion of the 28-day exposure period. Likewise, the serial dilutions used to assess populations of acid producing bacteria (APB) will change from an orange-red to a yellow color if "positive" for the presence of APBs. However, the tests for APBs are based on a 14-day exposure period. These changes in the color for the solutions, which are used to assess SRB or APB populations, are illustrated in Figure 21 below.

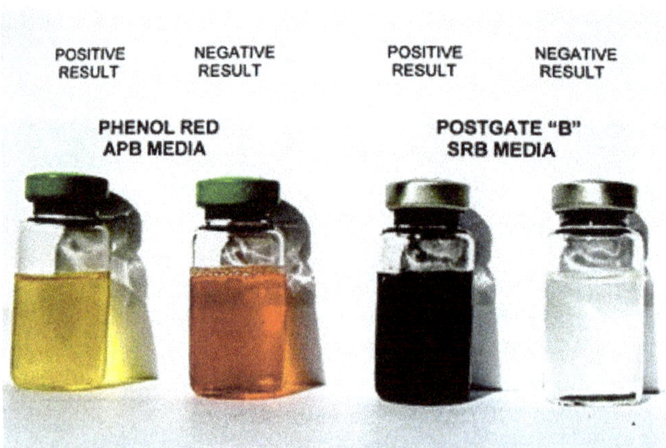

Figure 21 Bottles Turn from Clear to Black for Positive Indication of SRBs[28]

Typical industry jargon refers to the number of bottles that turn. In actuality it is recognized that the bottles will not change color–just the contents, and that is based on the microbiological populations and their consumption of the nutrients within each of the bottles.

The number of bottles or vials that change color provides the measure for the number of either the SRB or the APB microorganisms in the original sample. If one vial changes, the original sample has between 1–10 microorganisms/mL. If two vials change, the original sample had between 10–100 microorganisms/mL. Table 3 provides the estimation for the number of bacteria in the original 1 mL sample:

Table 3 Serial Dilution[47]

Vial No.	Number of Bacteria in Original Water Sample (1 mL)
1	1-10
2	10-100
3	100-1,000
4	1,000-10,000
5	10,000-100,000
6	100,000-1,000,000

Note that the presence of either SRBs or APBs does not mean that there is active microbiologically influenced corrosion (MIC) in the system. Instead, the results only confirm the presence of certain populations of those organisms, which could potentially attach themselves to the pipe wall, establish colonies, and ultimately cause MIC if left unchecked. However, lower populations of the microorganisms would also suggest a lower probability for MIC.

As applied to the formation of black powder, lower concentrations of SRBs means that a lower volume of H_2S could be generated through their metabolism. That means there would be less H_2S to react with the iron in steel pipelines or process equipment, and thus, the lower the rate for formation of iron sulfide, i.e., black powder.

5.3.2 Adenosine Phosphosulfate (APS) Reductase to Quantify SRBs

A second technique commonly used to quantify populations of sulfate reducing bacteria is the adenosine phosphosulfate (APS) reductase technique, which measures the concentration of a specific enzyme specifically associated with SRB. Field test kits are commercially available, and start by exposing the test sample to small particles containing antibodies that capture the APS reductase enzyme. Those enzymes are then exposed to a specific chemical indicator. The resultant color change is directly proportional to the concentration of the APS reductase in the sample. Thus, the population of SRB in the test sample can be rapidly estimated.

This particular test is unique to, and only capable for, detecting and quantifying SRB. The tests are designed to be conducted in the field, and provide results quickly, typically within approximately an hour, rather than the 28-day period required for the serial dilution technique. However, the APS reductase technique is several orders of magnitude less sensitive than the serial dilution techniques, which was described in Section 5.3.1 above.

Note that the APS reductase test for SRBs cannot discriminate between live bacteria and those that were freshly killed, such as from a biocide treatment. Accordingly, the APS reductase techniques should not be used to assess SRB populations on samples collected immediately after a biocide treatment.

When trying to assess the effectiveness of a biocide treatment for treating a pipeline, the best results would be from serial dilution tests, using samples collected very shortly before and after the biocide treatment. However, if answers are needed more quickly, consider using the APS reductase technique. Quantify populations using fluid samples collected immediately before the biocide treatment, as well as samples collected 24–48 hours after the biocide treatment.

There is a short shelf life for the chemicals used in the APS reductase tests. The shelf life is even shorter in warm climates. Hence, it is recommended to only order the number of test kits needed for the survey, and coordinate delivery of new test kits to coincide with the schedule for sample collection and field tests.

5.3.3 Additional Techniques That May be Used to Quantify Microbiological Populations

There are other techniques, which can be used to quantify microbiological populations in fluid samples. New techniques will undoubtedly be developed as the industry learns more about microorganisms and any affect they may have related to corrosion mechanisms. The following subsections will identify three additional techniques to assess microbiological populations. However, those tests are *not* specific to SRBs, but instead quantify *all* microorganisms. As such, these would primarily be used for specialty studies. These are encapsulated below to provide awareness of the methodology. If such specialized analyses are warranted, it may be best to delegate such analyses to company specialists or consultants.

5.3.3.1 ATP Photometry

Enzyme immunoassay methods can be used to quantify the presence of the enzyme adenosine triphosphate (ATP) found in all living cells. In concept, the test is very simple–the concentration of ATP in the field sample is proportional to the number of living bacteria in the sample. The ATP is quantified, using photometers that measure the light emitted when the ATP reacts with an enzyme. Commercial kits are available.

Note that the ATP photometry test is reasonably easy to conduct, and provides results quickly, but does not distinguish between different kinds of bacteria. Thus, for example, SRBs and APBs could be quantified together. Also, the presence of chemicals, such as biocides, oxygen scavengers, and H_2S, can interfere with and distort the results. Accordingly, a different test may be employed–hydrogenase measurements.

5.3.3.2 Hydrogenase Measurements

Hydrogenase is an enzyme produced by bacteria that use hydrogen as an energy source. In this method, the bacteria populations are quantified by adding an enzyme to the sample solution to extract the hydrogenase. That extracted hydrogenase is then preserved in a solution to maintain enzyme activity, and is then placed in a reaction chamber, where hydrogen is introduced. The hydrogen oxidation is indicated by a color change, and can take between 30 min and 4 h to develop. Thus the population of bacteria is determined from the reaction time and the associated color change.

5.3.3.3 Fluorescence Microscopy

Fluorescence microscopy is still another technique to quantify microbiological populations from a sample. This particular technique provides a direct, visual counting of bacteria, using an epifluorescence microscope. The technique calls for the samples to be "fixed" to kill the bacteria in formalin, and the samples are then stained with a dye, such as fluorescein isothiocyanate (FITC), which fluoresces under ultraviolet light. Operator skill is essential for valid results. As with the other test methods noted in Section 5.3.3, fluorescence microscopy does not distinguish between different types of bacteria (SRBs and APBs), and also does not distinguish between living or dead bacteria.

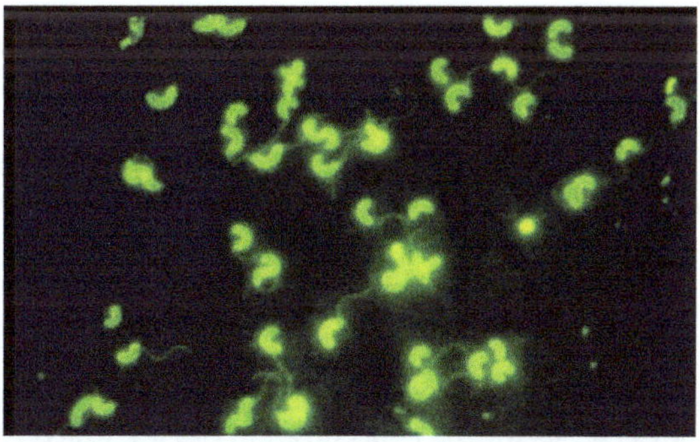

Figure 22 Fluorescence Photomicrography to Quantify Microbiological Populations

5.4 Collect Samples for Analysis of Solids

As part of the routine surveillance of gathering and production systems, samples of any solids should be collected and analyzed at regular intervals, such that potential hazards can be identified. Any significant changes in the analyses results should be examined.

Consider for example solids collected on valves or orifice plates, which might be taken out of service if production is dropping. Is there a buildup of waxy material, which is choking flow? If so, consult with the technical representatives of the chemical suppliers, the company specialists, or the producer's specialists or consultants to determine whether it might be appropriate to consider chemical treatments, such as through the application of paraffin inhibitors. Similarly, consider the possible use of scale inhibitors if scale is accumulating on the valves or orifice plates. What action should follow if there is a significant volume of iron sulfide obtained at the pig receivers whenever the pipelines are cleaned using maintenance pigging?

This section focuses on test methodologies typically used to determine the identity of deposits of black powder, such that the appropriate remedial measures can be used to either control the formation of such products, or assist in their removal from the system. The key is to look at all monitoring and sample analyses reports and look for consistencies. Look for results from analysis of materials removed when cleaning individual lines, but also look for significant changes in composition or concentration.

Note that this book does not provide an exhaustive listing of the analytical techniques. Instead, it identifies the many of the most commonly used analytical techniques used, not only for analysis of solids removed from the system, but also can be used for analysis of black powder.

5.4.1 Review Historical Knowledge of the Systems When Planning Sample Collection

Black powder is typically first detected when deposits are removed from the pipeline through maintenance pigging. However, the presence of black powder should not be a surprise. Records should be available from upstream producers or processors, which characterize the products to be transported. Their records and analytical reports may also raise the possibility for the formation of black powder, such as confirmation for the presence of H_2S, a precursor for the formation of iron sulfides.

If the pipeline will transport product from a new well, the producer should know the composition of that product, based on the initial well tests. They should have a gas analysis that includes checks for the concentration of H_2S, which represents a safety hazard, as well as CO_2, which relates to the BTU value of the produced gas. The concentrations of H_2S and CO_2 are important, because they relate to the potential for internal corrosion if an electrolyte (liquid water) is present. The producer would also know the water cut, as well as the initial production or flow rates. High flow rates may suggest the transportation of formation fines, such as sands, which indicate a potential concern for erosion or erosion-corrosion. Also check the histories associated with the production from any other wells from that same reservoir.

If the product to be transported is from a processing facility, obtain a copy of the specifications for that particular product, and focus on those that may pose an elevated threat to the system integrity. Check the operating temperatures and pressures and flow rates, versus the nominal design specifications. Samples are often collected off the filters or in the separators.

If the pipeline already has an operational history, one of the first questions is "When was this particular pipeline most recently pigged?" "What were the results from that pigging?" The focus is not only knowing the ***volume*** of material removed, but also the ***composition*** of that material. Identify the timeframe for when the pipeline was most recently pigged, such as 90 days earlier. Search the records to find the volume of debris removed in the most recent pig run, such as whether the volume removed is as small as two liters or perhaps as large as two barrels. Also check to determine what type of pig was used to remove that debris. Will multiple pig runs be necessary to completely clean the pipeline?

Ask whether there were any liquids, such as liquid hydrocarbons or produced/condensed water, and also, what were the results from previous analyses of gasses for the concentration of any hydrogen sulfide (H_2S)? Changes in the concentrations for each of the parameters being tracked will relate to the propensity for internal corrosion and system integrity.

Corrosion control professionals should check with operations or support staff, who may be more familiar with specific process vessels, to determine whether there are potential concerns. For example, although a gas stream may contain only 0.5 ppm H_2S, if the downstream slug catcher is partially filled with iron sulfide, then sample collection and analysis should include field tests to quantify the present concentration of H_2S and determine whether there may be significant changes in the concentrations of the parameters being measured.

Also beware of the possibility of naturally occurring radioactive material (NORM) in the samples. The preplanning for sample collection should also include a check to determine whether additional personal protective equipment (PPE) is warranted.

5.4.2 Field Sample Collection

Samples, which are collected at sample points or pig traps along a pipeline system or in process equipment, will typically be subject to (a) time-sensitive tests conducted in the field, and (b) more precise laboratory analyses.

- If the samples are liquids, then measurements such as pH should be conducted in the field as quickly as possible, because of the tendency for dissolved gasses to "out gas" and change the pH.
- Hydronium paper should also be used in the field to determine whether water is present within the liquid samples. The hydronium paper would turn from a bright white color to a brownish or purple color if water is present.
- For solid samples, such as black powder, take a clean scoop and place a small quantity in a clear plastic sandwich bag. Then move a magnet on the outside of the bag. Any magnetic material present in the bag will respond to the magnetic field, and would indicate the possible presence of iron oxides and iron sulfides.
- A second test for solid samples would be to take a very small volume of the black powder sample and place the sample in a small glass vial. Then add a couple of drops of hydrochloric acid onto that powder. Dangle a small strip of *lead acetate paper* from the rim of the glass vial, and quickly close the lid. If the sample contains FeS, the hydrochloric acid will react and release H_2S, which is well known for its distinct smell like rotten eggs. That release of H_2S will change the color of the white *lead acetate paper* to a brownish or purple color, providing a positive indication for the presence of sulfides (see Section 5.1.2.1).

NOTE: SAFETY FIRST IN SAMPLING

Black powder can be pyrophoric. That means that it can burst into flames without needing an external spark. Have water readily available whenever opening pig receivers or cartridge filter assemblies. Wet the interior of the pig receivers as soon as practical, and/or wet the filter cartridges as soon as they are removed from service.

Water serves as a heat sink and temporarily prevents spontaneous combustion. Keep the black powder wet until proper disposal. Note that the water does not cause a chemical change. If allowed to dry, the pyrophoric nature of the material may again be manifested.

5.4.3 Laboratory Tests to Identity Specific Types of Black Powder

Sections 5.4 and 5.5 address the type of laboratory tests that should be requested to confirm whether the product removed from the system is a combination of iron oxide and iron sulfides or some other product. This section also identifies analyses that can be used to identify specific forms of iron sulfide. These include a brief overview of the analytical technique to provide a basic understanding for the analyses, the limitations associated with the techniques, and how the results should be interpreted. The Case Study in Section 5.5 provides an example for the methodology used to analyze a sample described as "black powder," which was obtained from a new gas processing facility very shortly after it became operational.

5.4.3.1 Identify Elements First, Then Analyze to Identify Specific Compounds

One of the first steps is to identify the elements that are present in a sample. X-ray Fluorescence (XRF) is often one of the first analyses ordered, because it can identify numerous elements–not just the anticipated elements. Although there are stand-alone XRF units, XRF studies are commonly conducted using scanning electron microscopes (SEM) that have the XRF functionality. This is commonly called Energy Dispersive Spectroscopy (EDS). However, many SEMs with EDS/XRF functionality may not be able to detect lighter elements, such as carbon. In that case, be sure to supplement any EDS/XRF studies with additional chemical analyses to detect and quantify the presence of carbon, hydrogen, nitrogen, and sulfur (CHNS). CHNS analyses are commonly used to identify those elements in a sample, and are discussed in Section 5.5.3.

These analyses would typically be followed by x-ray diffraction (XRD) studies, which identify the specific forms of the iron sulfide or iron oxides in a sample. If other elements are identified in the analyses, additional analyses may be warranted.

Once the composition has been identified and potential safety/handling hazards have been addressed, future analyses of pigging samples would typically be conducted only periodically, or whenever there appears to be a change in the composition or volume of material removed during any individual pigging run. This is a practical consideration for field operations and the support requirements.

5.4.3.2 Expected Compositions Guide Selection of Analytical Techniques

Any analytical tests undertaken should be based in part on expected results. When analyzing a sample removed from a pipeline or process equipment, first consider the metallurgy of the pipeline or process equipment and that the sample is likely to be corrosion products from the base materials. Next, consider the possibility that the sample is a degradation product from the product or products being transported, or it may be a contamination.

- If the piping or process equipment was fabricated from stainless steel, look for the presence of iron, chrome, and nickel. Disproportionate ratios of these elements, relative to the ratio of their concentrations in the base metal, could be indicative of specific corrosion mechanisms.
- Metal samples obtained from carbon steel piping or process equipment may also provide an insight into the corrosion mechanisms that may be occurring. Besides for iron, look for the presence of manganese, a common element added to steels to increase strength and hardness. Also, check for the presence of carbon, which is commonly added to strengthen steels. Compare the relative concentration of iron to that of other elements within the alloy, such as man-

ganese in carbon steels or chromium and nickel for stainless steels. That data may provide some insight into the active corrosion mechanisms.

- Compositional and microstructural analyses can confirm that the proper materials were used in the original fabrication. Also, note that some analytical tools are not sufficiently sensitive to be able to detect carbon. Hence, complementary analytical techniques are typically employed, as described below.
- Next, consider the composition of the product or products being transported. For crude oil and natural gathering and processing systems, expect hydrocarbons (carbon, hydrogen, and oxygen), as well as produced or condensed water (hydrogen and oxygen). There could also be contaminants, such as H_2S (hydrogen and sulfur) from any produced or dissolved gasses, or salts. For samples collected from offshore systems, look for the presence of magnesium, because that element is most commonly found in seawater. Sand or dust, typically in the form of sodium silicates, may be swept into the pipeline systems from production wells, or could have been introduced during the initial hydrostatic pressure tests following the initial construction of the pipeline.

Look at the elemental composition and chemical structure of the samples, where the components may have come from, and use that data when evaluating threats posed to system integrity.

5.5 Selecting Analytical Techniques for Lab Personnel to Use When Assessing Solids

This section highlights the most commonly used techniques for analyzing solids. There will be a brief overview of the analytical techniques the laboratory personnel employ and the type of data generated. For example, XRF and SEM-EDS analysis would be used to identify the elements present in a sample. Those analytical techniques identify the presence of iron and sulfur, but would not provide any information regarding whether that sample is an iron sulfide, or perhaps an iron sulfate. However, other complementary analyses, such as X-Ray Diffraction (XRD), would be used to identify the specific compounds.

The primary message is that multiple techniques are typically best for identifying solids samples and making sure nothing significant or unexpected is overlooked. Use XRF or SEM-EDS to identify the elements present in samples of solids, including elements that were not expected. Samples of black powder would be expected to have strong signals for iron, oxygen, and sulfur, corresponding to the expected iron oxides and iron sulfides. However, if other elements are reported, conduct follow-up discussions with the laboratory analyst to identify those specific compounds. For example, are those elements representative of components found in compressor lubrication oil? Alternately, be aware of other components found in hydrocarbon production. For example, if there were strong indications for barium and sulfur, there is a possibility that the system may be experiencing barium sulfate scale depositions. That is not a concern related to the presence of black powder, but would need to be addressed, because that particular scale is very tenacious and difficult to remove. Similarly, if the elemental analysis suggests the presence of aluminum (Al), zinc (Zn), and silicon (Si), then look for particles from the desiccant used in gas processing. Again, the presence of Al, Zn, and Si does not relate to concerns associated with black powder, but are nevertheless related to gas processing and dehydration operations, which may need to be addressed. The follow-up or extended analyses beyond XRF or SEM/EDS would be conducted on an as-needed basis, using instrumentation like that presented below in Sections 5.5.1–5.5.3.

5.5.1 X-Ray Fluorescence (XRF)

This is an analytical technique to identify the elements that are present in a sample. X-rays are focused onto the sample, which excites the electrons orbiting the central nucleus of the atoms. When electrons on an inner shell are excited and jump to an outer shell or are ejected from the atom, one of the remaining electrons from an outer shell will jump to the lower shell, fluorescing (Figure 23), or emitting a characteristic amount of energy (Figure 24). The XRF spectrometers detect these emissions, and characterize the elemental composition. Additional information on the XRF technique is readily available through the Internet.[48]

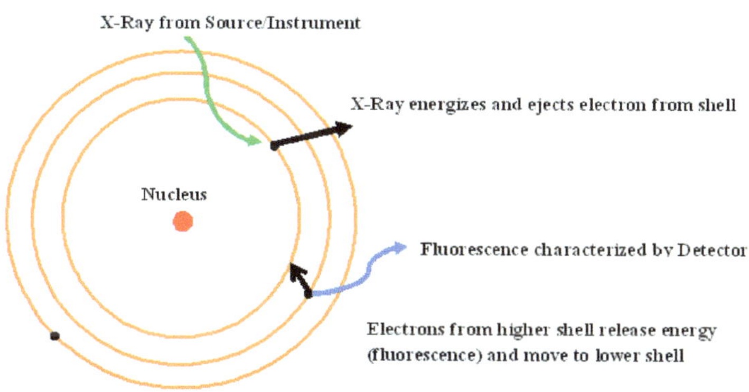

Figure 23 X-Ray Fluorescence—How it Works

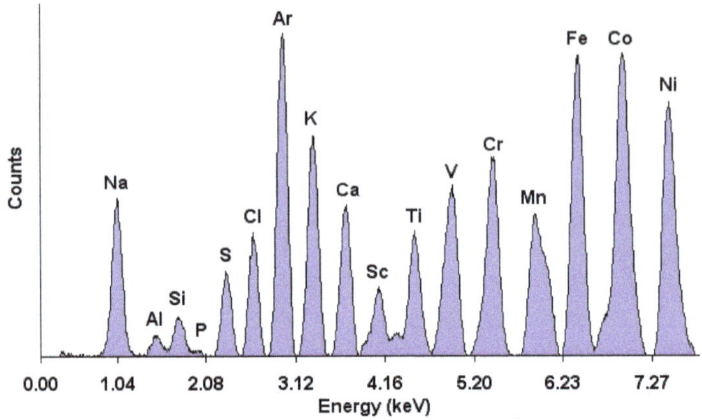

Figure 24 Typical Spectra from XRF, Identifying Elements by Their Energy Releases[49]
Image courtesy of © 2017 AMPTEK, Inc. All rights reserved.

X-ray spectrometers are commercially available as standalone units, or as additional modules for other analytical instruments, such as Scanning Electron Microscopes. SEMs having Energy Dispersive Spectrometry (EDX or EDS) capability can also identify elements in a powder. Although theoretically possible to detect elements as light as beryllium (Be), a more practical detection threshold is that of sodium (Na), because of low x-ray yields and instrument limitations.

Some, but not all SEM-EDS units, have the capability to detect carbon, which will be discussed in Section 5.6. Hence, other techniques are generally used to detect and quantify lighter elements, such as hydrogen, carbon, nitrogen, and oxygen.

XRF Used to Identify Elemental Composition

- Iron
- Chlorine
- Sodium
- Chromium
- Silicon
- Magnesium

If other elements are detected, investigate to confirm their presence and origin (source) of those species. This is for assessing potential threats to system integrity.

5.5.2 X-Ray Diffraction to Identify Structures from Their Crystalline Structure

Base metals, such as steel, have a crystalline structure, as do corrosion products. The repeating crystalline structure makes it possible to use x-ray diffraction to identify the specific compositions of samples collected for analysis. In this technique, a beam of incoming x-rays will bounce off of the crystal structure in many specific directions. The angles of those diffracted x-rays are measured, along with the intensity of those signals, relative to the intensity of the incident beam. This makes it possible to produce an image of the structure being analyzed. Researchers have already developed extensive libraries for the diffraction patterns of known structures. Hence, modern x-ray diffraction software matches the analytical results from the sample to the patterns in the library, thereby providing the identity of the sample.

Figure 25 illustrates the generation of an x-ray diffraction pattern, while Figure 26 illustrates the analysis of incident and scattered x-rays used to identify specific crystalline structures.

Scanning electron microscopes having the capability to detect and measure electron backscatter are functionally equivalent to x-ray diffraction units.

The bottom line is that XRF and XRD studies complement each other to identify the iron oxide and iron sulfide structures characteristic of black powder, and thereby provide valuable results for corrosion control professionals or operations personnel.

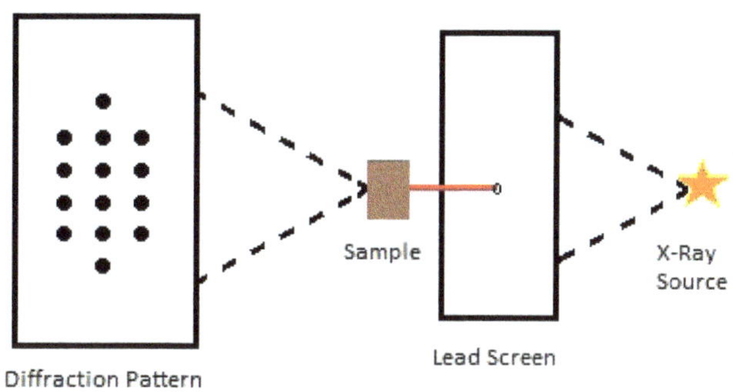

Figure 25 Illustration of X-Ray Diffraction
The spaces in the diffraction patterns are used to identify the composition of the sample. Further information can be obtained through an Internet search of "X-Ray Diffraction."[50]

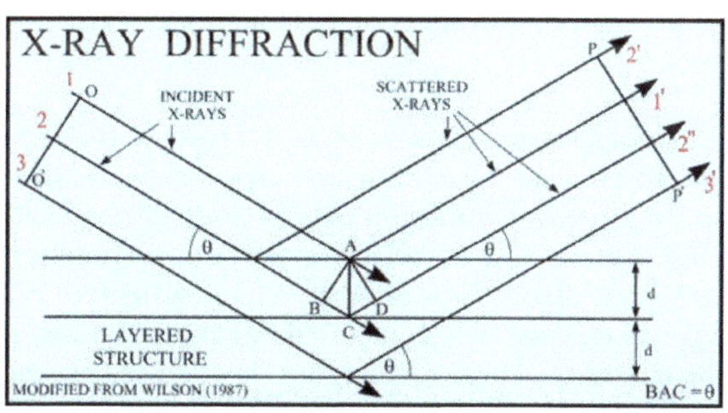

Figure 26 XRD for Analysis of Crystalline Structure of Sample[51]
XRD processing software has extensive libraries for matching patterns to identify samples.

Although corrosion control professionals or operations personnel may collect the samples, laboratory analyses should be conducted by trained laboratory personnel to identify and characterize the material removed from pipelines or processing equipment and assess the threats to system integrity.

XRD analyses are excellent for identifying the specific crystalline structures, such as

- Iron Sulfide
- Iron Carbonate
- Iron Oxides
- Other Compounds

5.5.3 CHNS Analysis and Wet Chemistry Techniques for Sulfides and Carbonates

Many scanning electron microscopes having energy dispersive spectrometry (SEM-EDS) capability may not be able to detect lighter elements such as carbon. As such, when analyzing samples, it is commonplace to request analysis for carbon, hydrogen, nitrogen, and sulfur (CHNS), which are commonly found in samples collected from crude oil and natural gas gathering and processing systems. The elements can be analyzed through specialized instruments that identify all elements simultaneously, or through wet chemistry techniques. Here, "wet chemistry" refers to the classical methods for analyzing components of a solution through multiple step processes, including titrations to separate the components of interest and quantify the components of interest.

The CHNS and wet chemistry analysis for sulfur can provide a measure for sulfates or sulfides present. Similarly, measures for carbon may indicate the concentration of carbonates, or perhaps something else. Section 5.6 will present an example for analysis of a black powder problem reported from a new gas processing facility. As discussed in that chapter, the CHNS analysis of the sample yielded something very significant: the observation of very high concentrations of carbon, which led to an analysis and evaluation of the gas processing itself. Results from CHNS analysis, whether through instruments or wet chemistry, can be very significant in root-cause analyses.

CHNS analyzers use high temperature (1830 °F [1000 °C]) combustion in oxygen rich environments to convert the carbon to carbon dioxide, the hydrogen to water, nitrogen to nitrogen gas or nitrogen oxides, and sulfur to sulfur dioxide. The combustion products are swept out of the combustion chamber by an inert gas, such as helium. A high purity copper strip, which is heated to 1110 °F (600 °C) removes any oxygen not previously consumed, and converts any nitrogen oxide to nitrogen gas. The gasses can then be analyzed through several techniques, such as gas chromatography separation and thermal conductivity detectors.

The CHNS analyses provide information related to the carbon content of a sample, as well as that for the hydrogen, nitrogen, and sulfur content. XRF analysis of black powder samples would be expected to report the presence of iron and sulfur. Results are reported, based on the sum of the *detected* components such that the sum of those components equal 100%. However, most commercially available XRF or SEM EDS/XRF instruments are not able to detect carbon. Consequently, if components containing carbon are present, there may be a major discrepancy in results from the same sample, if the analyses were obtained from two different SEM EDS/XRF instruments: one that has the ability to detect carbon, and one that does not have that capability. This will be seen in an example in the next section, which describes the analytical approach used to assess a sample of black powder obtained from a facility. Accordingly, it is recommended to supplement XRF or SEM EDS/XRF analyses with wet chemical analyses, such as CHNS, for components that may not be detected when the available analytical equipment may have limited capabilities. Verify the analytical instruments used have the capabilities to detect and quantify the chemical elements or anions of interest.

CHNS Analysis Can Quantify Presence of

- Carbon
- Hydrogen
- Nitrogen
- Sulfur

Wet Chemistry Can Quantify Presence of

- Sulfides
- Carbonates

5.6 Case Study: Illustrating Methodology to Assess Powder Sample from a Facility

This case study is based on a black powder problem encountered at startup of a new gas processing facility in western Colorado.[31] Although the facility reported a major problem with a dry, black powder, which clogged the in-line filters and piping in the facility, as well as the export pipeline, the actual problem was not associated with iron sulfides or iron oxides. This case study illustrates the methodology by which that powder was analyzed and how the root cause of the true black powder problem was determined. Based on that analysis, it was necessary to make changes to the gas processing in order to halt the generation of the product initially described as black powder. A detailed description of the problem, the analytical methodology to identify the root cause, and the remedial efforts was previously published.[31] This section will describe the analytical steps undertaken, because it is germane to any investigation of black powder.

The discussion will start with an overview of the gas processing facility, in that it helps identify sources of elements or compounds, which investigators may want to consider during the initial investigations to characterize the powder and its possible origin. Natural gas entering this facility first passes through inlet filters, and then flows through an amine unit to remove any carbon dioxide (CO_2) and any hydrogen sulfide gas (H_2S). The flow is then directed through one or two of the three inlet dehydrators, and then flows to the cryogenic separators, which separates the methane, ethane, propane, and higher hydrocarbon components. The dehydrators are packed with industry standard sodium potassium aluminosilicate (73–85%) and magnesium aluminosilicate clay (15–27%), or sodium aluminosilicate (73–85%) and magnesium aluminosilicate (15–25%) desiccants. The composition of these desiccants is significant to this discussion, because the analysts looked for the presence of the elements associated with each dessicant to be within the samples of the dry powder, which were collected downstream of the dehydrators. At most, only trace amounts of aluminum, silicon, magnesium, or potassium were detected. Hence, it was possible to eliminate a failure of the dehydrator packing as a source for the reported black powder.

This particular facility processes approximately 490 million standard cubic feet per day (MMscfd) (13.875 MM m^3/d), with approximately 450 MMscfd being routed through one or two of the dehydrators. The remaining 40 MMscfd (1.133 MM m^3/d) is routed through the Waste Heat Recovery Units (WHRU) to heat that gas stream, such that it can flow through and regenerate the third dehydrator by driving off the moisture. The gas stream used to regenerate the dehydrators is referred to as "regen gas." It passes through a heat exchanger, and is warmed by the hot exhaust of the facility's turbine-driven compressors.

Figure 27 illustrates the process flow and where the product initially described as black powder was found. Figure 28 illustrates fine, loosely adherent black-colored powder, which was displaced from the outlet header of the WHRU using a fan normally used for room ventilation. Figure 29 illustrates inlet gas dust filters downstream of the dehydrators and upstream of the cryogenic separators. The pressure differentials built quickly, necessitating frequent and expensive change-outs. Such buildup of differential pressures also served as a notice to operations personnel of a production or processing problem.

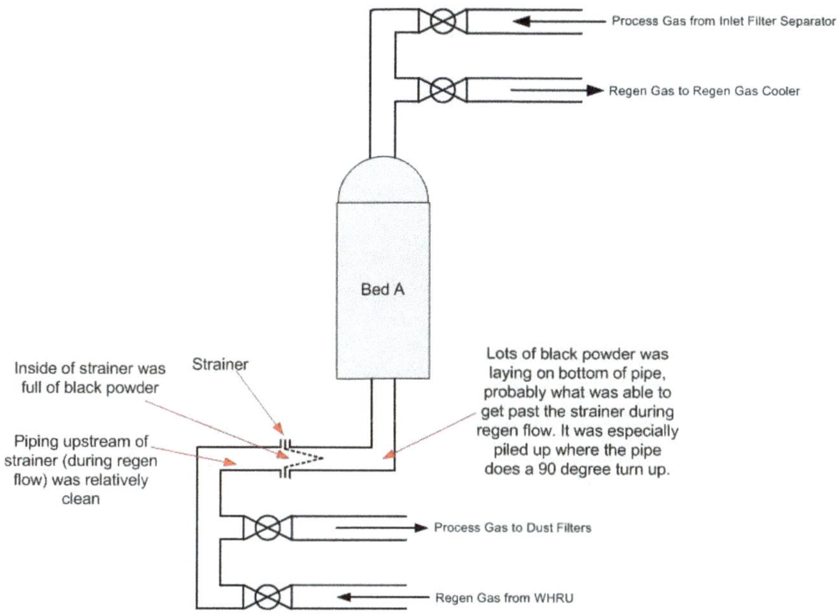

Figure 27 Process Diagram Illustrating Where "Black Powder" Was Found[31]

(A) (B)

Figure 28 Piping from Outlet Header of WHRU Filled with "Black Powder"[31]

(A) Fine, loosely adherent powder is displaced from the top of a buildup or bed of particles by an air fan at the opposite end of the pipe segment. (B) Hard particulates, presumably (Fe_3C), ricocheting off the bottom of the blind flange, located on the far left. Analyses of the powder confirmed the presence of Fe_3C.

Figure 29 Inlet Gas Dust Filters Became Clogged with Fine Powder[31]
This caused differential pressure to increase, necessitating frequent change-outs.

5.6.1 SEM-EDS Analysis Used for Elemental Composition

Table 4 illustrates results from SEM-EDS for identifying the elements that were present, namely those that that particular SEM-EDS instrument was capable of detecting.

- The software associated with the SEM-EDS instrument makes an assumption for the most probable compound associated with the elements detected, and reported those results, such that the sum of the weight percentages totals 100%, or normalized.
- Very few SEM-EDS instruments are capable of detecting and reporting light elements, such as carbon.

The 8-19-09 Dust Filter A sample was analyzed by two different SEM-EDS instruments, one of which had the capability to detect elements as light as carbon, and one that could not detect the presence of carbon. The two very different results are shown in Table 4 below. As is customary for such analyses, the sum of the components for each analysis totals 100%, sometimes called the "normalizing of results." Looking only at the results for the concentration of iron, the instrument that was not capable of detecting carbon, reported 83.564% iron. In contrast, the instrument that could detect carbon, reported 31.380% iron, and 64.270% carbon for the same sample. The message is that even if one analytical instrument does not detect a potentially significant element or component in a sample, verify such results through the use of another instrument or analytical tool.

The full SEM-EDS analyses are presented in Table 4. (The red values here and in Table 5 document the presence of carbon compounds.) There were a number of different elements detected, which would not normally be expected in samples collected from the natural gas process streams. These included sodium, aluminum, potassium, and silicon. However, those elements are components of the desiccant beads used in the dehydrators, and that shows the value of reviewing the process diagrams. (Also, the original analyses included a strong Kα spectral peak, which could be attributed to either sulfur or molybdenum, because they are very close in energy spectra. However, the concentrations for the sulfur (or molybdenum) were significantly lower than the 20% concentration, relative to the concentration of iron that would be expected if the powder contained a mixture of iron sulfide and iron oxide. Hence, other techniques were necessary to understand the mechanism at play.

Table 4 Results from SEM-EDS Elemental Analysis of Powder Sample

		First Sample (8-12-09)	Dust Filter A (8-19-09)	Dust Filter A (8-19-09)	WHRU B Inlet Header (9-29-09)	WHRU B Outlet Header (9-29-09)
Parameter	Assumed	Wt. % *	Wt. % *	Wt. % ##	Wt. % *	Wt. % *
Sodium	Na_2O	0.470				
Aluminum	$Al_2O_3SiO_2$	1.530	6.499	0.350	0.560	4.502
Silicon	SiO_2	7.760	3.412	0.760	0.528	1.266
Phosphorous	P_2O_3	0.260		1.880		
Sulfur/ Molybdenum	S/Mo	0.380	1.628	0.400	0.232	0.290
Potassium	K_2O	0.210			0.055	
Calcium	CaO	2.920	2.600	0.170	0.196	0.557
Chromium	Cr_2O_3	1.360	2.089	0.800	0.383	2.130
Iron	Fe_2O_3/Fe_3O_4	83.990	83.564	31.380	96.719	86.823
Copper	Cu_2O	1.110			0.317	0.664
Manganese	Mn	-	0.208			
Chloride	Cl				0.100	0.231
Zinc	Zn				0.911	3.361
Carbon	C	-	-	64.270		
		*This SEM/ EDS not able to detect Carbon	*This SEM/ EDS not able to detect Carbon	##This SEM/EDS can detect Carbon	*This SEM/ EDS not able to detect Carbon	*This SEM/ EDS not able to detect Carbon

SEM/EDS results are "normalized" such that the sum of all observed components totals 100%. The results from analyses of the same sample can be completely different, depending whether the SEM/EDS has the capability to detect the presence of carbon. This is illustrated in Table 4 by the analysis of the 8-19-09 sample, with the red font color denoting the presence of carbon compounds. Know the limitations of analytical instruments when evaluating results.

Next, XRD analysis of the powder sample is presented in Table 5.

Table 5 Typical Results from X-Ray Diffraction Analysis of Powder Sample

		First Sample (8-12-09)	Dust Filter A (9-11-09)	Dust Filter A (9-29-09)	WHRU B Inlet Line (9-29-09)	WHRU B Inlet Header (9-29-09)	WHRU B Outlet Header (9-29-09)
Mineral	Chemical Formula	(%)	(%)	(%)	(%)	(%)	(%)
alpha-Quartz	SiO_2	1					
beta Quartz	SiO_2	7	3		2		
Cristobalite	SiO_2	Tr	Tr				
Calcite	$CaCO_3$	2			Tr		
Magnetite	alpha Fe_3O_4	66	18	15	19	75	6
Hematite	alpha Fe_2O_3	5	1	Tr	1	11	
Cementite	Fe_3C	17	28	7	8	1	21
Goethite	alpha FeOOH	1					
Illite	$KAl_2(Si_3AlO_{10})(OH)_2$	1					
Desiccant	$NaAlSO_4)12\text{-}27\ H_2O$			2			
alpha Iron	Fe						
Wustite	FeO	3		5		13	
meta Anthracite	C		35	75	70		73
Amorphous	C		15				

Results from XRD analyses of powder samples may be drastically different, depending on the minerals analyzed. Compare the WHRU B inlet header samples collected on 9-29-09. One sample had 75% magnetite reported when analysis did not report anthracite, while the second sample from the inlet header line reported 19% magnetite and 70% anthracite. The red font color highlights analyses that specifically focused on identifying and quantifying anthracite and amorphous carbon.

Each of the XRD sample analyses results were normalized to 100%. These analyses report iron oxides [magnetite (Fe_3O_4), hematite (Fe_2O_3) and wustite (FeO)], iron carbide [cementite (Fe_3C)], the desiccants, and silicon dioxides, and most significantly carbon (anthracite and amorphous). There was no evidence of any sulfur-bearing compounds, such as iron sulfides, which the laboratory equipment would have been able to detect if it was present. These results confirmed that this particular powder, although dark black and in the form of a fine powder, was not an iron sulfide.

Noting that there was *no* evidence of iron sulfide upstream of the inlet dehydrators, the root cause investigation focused on the process for heating a portion of the inlet stream (40 MMscfd (1.133 MM m^3/d)) to high temperatures and using that heated gas to regenerate the dehydrators, once they were water saturated. Again, the goal was to identify the root cause source for the observed "black powder," such that further generation of that powder could be eliminated.

The temperature of the turbine exhaust gas was measured as 1027 °F (553 °C) just downstream from the regeneration coils. Those temperatures were sufficiently high to allow the higher hydrocarbon contents of the inlet gas, namely the ethane, propane, butane, etc., to decompose into carbon and hydrogen. This was further confirmed by the collection and analysis of one special gas sample collected from the WHRU headers, which was found to contain 7% (mole) hydrogen gas (H_2).

Some of the elemental carbon generated would also be expected to react with the iron in the steel piping and headers to form the observed iron carbide cementite (Fe_3C), as identified in Table 5 above. Accordingly, the WHRU regeneration processes were changed, such that a portion of pure methane (CH_4) gas stream, which was obtained after the separation process, would be used to regenerate the dehydrators. Again, this change was necessary, because the temperatures of the turbine exhaust at the WHRU was measured at 1027 °F (553 °C), which is above the temperatures at which butane, propane, and ethane components of the inlet gas typically decompose into hydrogen and carbon. (The relative auto-ignition temperatures are approximately 761 °F (405 °C) for butane, 851 °F (455 °C) for propane, 959 °F (515 °C) for ethane, and 1112 °F (600 °C) for methane. These are the temperatures necessary to provide the activation energy for auto ignition in environments containing oxygen, if there was no external source for ignition (such as a flame or spark.) Thus, by changing the process, such that only pure, processed methane is used to regenerate the dehydrators, the generation of the anthracite charcoal was minimized, while still capturing and using the waste heat from the turbines.

This case study illustrates the systematic approach to collecting and analyzing samples, recognizing the limitations of analytical techniques, using results from multiple analytical techniques, and understanding the processes in the facilities. As a result of this analytical approach, the root cause of the generation of the charcoal, which was mistakenly called black powder, was identified, such that changes could be made to the operational processes to minimize further generation of that charcoal. Additionally, knowledge of the composition of that powder provided the direction for removing that powder from the system and cleaning the affected piping and process equipment.

Methodology to Confirm Identity of Black Powder

- Collect sample in "the field" (consider multiple samples to diagnose root cause and identify source)
- Document conditions
- Coordinate with analysts, such that the probable composition of sample is known, and the analytical techniques can be selected to ensure detection of the suspected components.
- Obtain Elemental Composition from X-Ray Fluorescence or Scanning Electron Microscope having Energy Dispersive Spectrophotometry
- Consult with analyst for additional analysis recommended to confirm identity of unusual elements identified in the analysis. Determine their significance.
- Consider X-Ray Diffraction to identify the specific composition of the solids. Once known, only periodically analyze to confirm no changes.

5.7 Time-Sensitive Analyses and Not Time-Critical Analyses

This section identifies analyses that should be conducted onsite, immediately after collection of the sample, and those analyses, which can be conducted at a later time, such as after the sample has been transported to a supporting laboratory. Here, "time sensitive" refers to analyses that must be conducted immediately after sample collection to remain representative of the conditions in the pipeline or process equipment. Otherwise, conditions such as pH or dissolved CO_2 may change, such that any remedial measures may not be as effective as originally intended.

The initial sampling and analysis is to identify conditions that may pose a safety or health hazard to those personnel collecting the samples. These include potential concerns for naturally occurring radioactive materials (NORM) if this is (a) the first sampling at a sample collection point, (b) the first maintenance pigging run, and (c) determining whether the samples may be pyrophoric.

If the black powder sampled or removed from the system includes NORM, any of those radioactive materials could pose a long-term health hazard if ingested into the lungs. Hence, that determination is a top priority for identifying the necessary personal protective equipment (PPE) for protecting personnel on-site.

Next, the black powder should be assessed quickly to determine whether it might be pyrophoric, suggesting the possibility of spontaneous combustion that would pose a safety hazard to personnel supporting the sample collection or maintenance pigging operations. Pyrophoric black powder is a form of iron sulfide, which may smolder or burst into flames if allowed to dry (which removes the heat sink) and when exposed to oxygen. Hence, the first tests are to determine whether the material is magnetic (iron), and then to determine whether that powder is in the form of a sulfide. The latter can be determined quickly by observing if a small sample fizzes and releases H_2S gas when wetted by a small drop of hydrochloric acid. The H_2S has a distinctive rotten egg smell, and causes lead acetate paper to turn from a bright white to a brown or purple color. Such positive indications dictate the need for precautions and that the material should be handled as if it is pyrophoric.

Once those health and safety hazards have been identified, additional sampling and field analyses focus on properties, which may change rapidly with time. For example, any carbon dioxide dissolved in a sample of produced water will come out of solution in minutes. Field tests are essential for measuring the pH of a solution, using pH paper, as well as measurements of the CO_2 content dissolved in the produced water. It is also best to prepare serial dilution studies for quantifying populations of SRBs or APBs as soon as possible after sample collection, before any changes occur to the pH or dissolved CO_2 of the solutions in which those microorganisms lived. Also, it is essential to place and keep those microorganisms in a dark environment, because direct sunlight can kill them.

Tests that are not time sensitive include XRF, XRD, and SEM/EDS analysis of solid samples for elemental analyses or structural information, which identifies specific compounds. Another test, which is not time sensitive, is particle size analysis, which can be for selecting the appropriate sized separation equipment, such as filters.

If tests are to be conducted to assess dissolved metals, such as dissolved iron, it is best to add approximately 5 mL of hydrochloric or nitric acid to 95 mL of the sample solution, so that the sample remains acidic, and the dissolved metal cations do not precipitate, or come out of solution. Again, this measure is to ensure the sample being analyzed in a laboratory remains representative of the system.

Field sampling and analysis (time sensitive) may include:

- **Field Analyses to detect Iron Sulfide**–Place sample in small glass vials, and place a drop of HCl acid onto the sample. Place a strip of lead acetate paper, so that it dangles or hangs from the top of the vial. Quickly close the cap of the vial. Check for fizzing and then a change in color for the lead acetate paper, as it turns from white to a brown or purple color, which is positive for the release of H_2S gas. (H_2S has a distinctive rotten egg smell, and can pose a health hazard.)
- **Magnetic material**–Hold a strong magnet outside the sample container to determine whether the material is magnetic.
- **Naturally Occurring Radioactive Materials** (NORM)–Determine if material is radioactive, which may pose a health hazard when handling.
- **Is material potentially pyrophoric?** Fill sample container, and keep it capped to retain moisture. (Also, filling and capping a container will minimize any oxygen ingress.)
- **Initiate serial dilution study to quantify population of SRBs or APBs** within aqueous samples or dormant on samples of solids.
- **Use pH paper and measure the pH of any samples of produced or condensed water.** Conduct as soon as practical, before outgassing of any CO_2 and H_2S changes the pH of the solution.
- **Collect any samples of produced/condensed water.** Fill container totally to minimize any gasses from going out of solution. Place a cap on container.
- **Measure the moisture content of gas** at the time of sampling.
- **Measure the CO_2 content of gas** at the time of sampling.
- **Measure the H_2S** within any produced/condensed water at the time of sampling.

Samples taken for laboratory analyses (generally not time sensitive) include:

- XRF to identify elements present, and XRD to identify chemical compounds.
- Chemical analyses of liquid samples to characterize anions and cations, including atomic absorption (AA) for elements and wet chemistry for other anions or cations.

  Consider adding 5 mL of a 1 Normal hydrochloric acid or nitric acid to 95 mL of an aqueous sample. This is to keep cations such as iron from precipitating or coming out of solutions before lab analyses.
- Particle Size analyses–Particles may appear as fine black powder, smoke like, or a wet, tar-like, slurry.
- Additional analyses, as warranted.

5.8 Summary of Analyses to be used to Characterize Samples of Black Powder

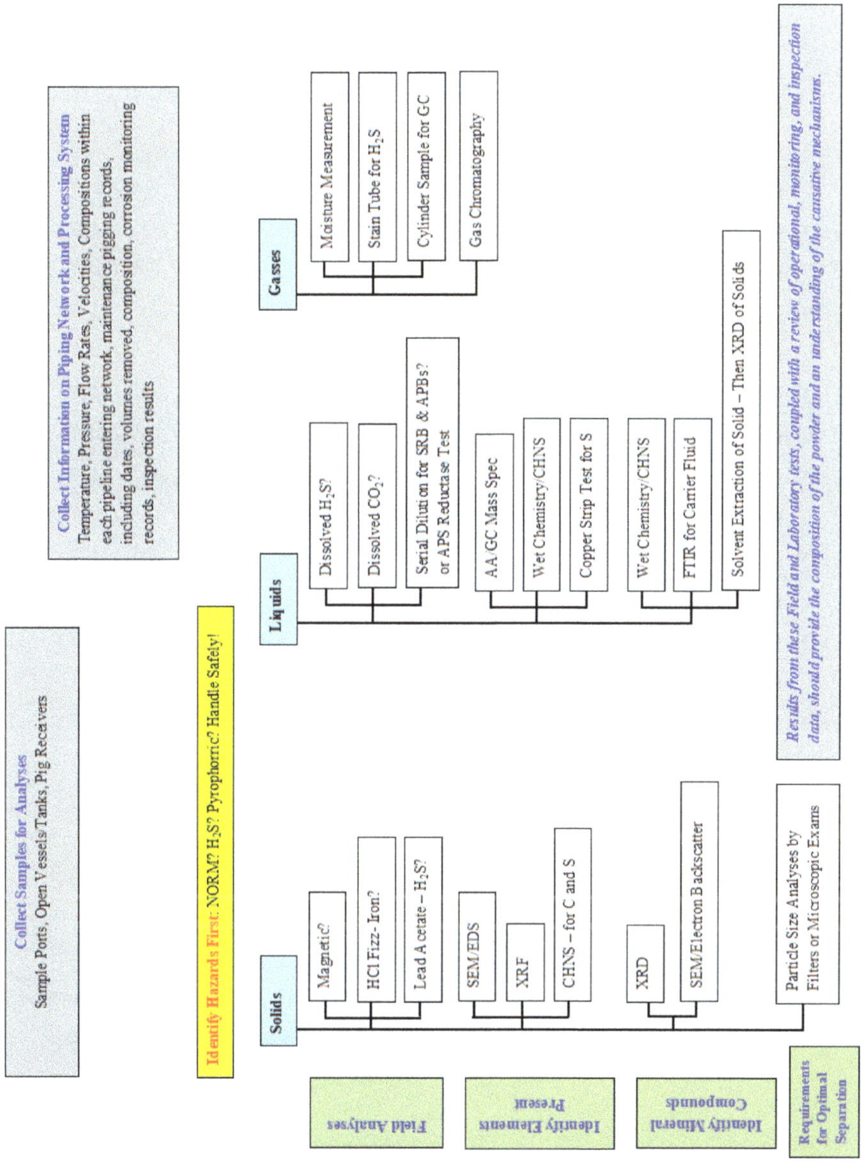

Figure 30 Recommended Analyses if Black Powder is Reported

6 Removing Black Powder by Pigging

Recall Figures 1 and 2: Figure 1 illustrates a polyurethane foam pig with a burn mark that occurred when a small patch of pyrophoric iron sulfide caught fire and melted the urethane foam. Figure 2 is a multidisc scraper disk pig used to rigorously clean a pipeline. After completing its run, it was hosed down with water to wash away any pyrophoric iron sulfide that remained on the pig, once it was retrieved from the pig receiver. Thus, any type of maintenance cleaning pig that displaces pyrophoric iron sulfide is susceptible to damage from pyrophoric iron sulfide until that black powder is removed. This chapter will present an overview of maintenance pigs, which can be used to clean pipelines, and will also set the stage for applying chemical products, which can be used to treat pyrophoric iron sulfide in the pipelines, thereby reducing safety and health hazards that corrosion control or operations personnel may encounter.

6.1 Maintenance Pigging—Types of Pigs

There are numerous manufacturers and service providers available to support maintenance pigging programs. Likewise there are a number of service providers available for supporting in-line inspections. Maintenance pigging is essential to keeping pipelines clean, and thereby increasing flow efficiencies. It is key to removing solids, which otherwise could plug filters at the customer receipt points. Maintenance pigging is also necessary for conducting good, valid in-line inspections of pipelines. Thus, an effective maintenance pigging program is an essential component of pipeline operations.

6.1.1 Foam Pigs

The primary roles for foam pigs are for pressure maintenance of pipelines by moving liquids, dehydrating pipelines, or applying chemical products to the interior of the pipelines. Foam pigs, such as those shown in Figure 31, are fabricated from polyurethane, and have the advantage that they are flexible and compressible or expandable. That means they can get through most pipeline restrictions, such as reductions in diameter or dents, including reduced port valves. It is also easy for operational or maintenance personnel to handle foam pigs.

The density of the polyurethane foam can vary from light to heavy. These pigs may also have an extra coat or jacket of dense polyurethane on the exterior surface, which provides resistance to wear, enabling them to traverse longer distances without getting torn, and provides longer service life.

Although good for forming a seal and helping drive liquids through a pipeline, they will flex and move over accumulations of corrosion products or debris, and consequently are best for dehydrating pipelines or applying liquid corrosion inhibitors, and are not used for removing solids from pipelines.

Figure 31 Foam Pigs and a Specialty Pig Used to Dehydrate a Pipeline

6.1.2 Batch Pigs for Chemical Applications—Typically Between Two Foam Pigs

Chemical products can be applied to the interior wall of a pipeline by placing the liquid solutions between two foam pigs, each of which forms a seal with the pipe wall. The two pigs function as a single unit, and are pushed or driven through the pipeline via the pressure differential. The chemical products may be biocides, corrosion inhibitors, or other specialty products. Later chapters will discuss the chemical cleaning of a pipeline to remove the black powder.

Occasionally, the length of the pig receiver is not sufficient to accommodate two batch pigs. Accordingly, pipeline operators often inject the chemical treatment directly into the pipeline, and follow it with a single foam pig, which displaces that liquid through the pipeline. Such chemical applications could be effective for treating the bottom of the pipeline, but does not ensure the 360° coverage to ensure application on the top of the pipeline. That is why chemical applications are best when applied between two pigs, and 360° coverage can be attained.

6.1.3 Spherical, or Ball, Pigs

Round, spherical pigs are commonly used to remove liquids, such as condensates or produced or condensed water, from pipelines. Spherical pigs are also able to negotiate or pass through short radius bends, enabling pigs from two different pipelines to be directed to a common pig receiver. That represents a cost savings in the pipeline construction. Although effective in displacing liquids, the dense spheres are more likely to compress loose deposition, rather than transporting those deposits to the pig receivers for removal. Figure 32 illustrates one such receiver, along with several spherical pigs in the foreground.

Figure 32 Launcher and Receiver for Spherical Pigs

Pipelines and launchers/receivers, which were designed and built for spherical pigs, are generally not able to accommodate in-line inspection vehicles. That is due to two considerations. First, inspection vehicles are typically longer, and may not be able to negotiate the tight radius bends. Second, the length of in-line inspection vehicles may exceed the distance between the pig receiver's ball valve and the trap door, meaning the valve could not be closed for isolation and safe removal of an in-line inspection vehicle.

6.1.4 Cleaning Pigs

The figures in the next few sections illustrate several different styles of cleaning pigs. Figure 33 illustrates a cleaning pig with brushes designed for passage through a single diameter pipeline. The brushes loosen black powder or corrosion product depositions from the interior pipe wall.

Figure 33 Brush Pig for Cleaning Pipelines
Photograph courtesy of TD Williamson

Cleaning Pigs with Attached Magnets
A variation for this same kind of pig could have a bank of magnets. Remember, iron oxide and iron sulfides are typically magnetic. Hence, iron oxides or sulfides, which are loosened by a brush pig run, could be removed by the subsequent running of a pig with magnets. Chapter 7 will present in-line inspections (ILI) using magnetic flux leakage (MFL) technology. That means the inspection vehicle has very strong magnets. In addition to inspecting pipelines, the MFL pigs have a reputation for being very effective in removing loosely adherent deposits, such as iron oxides or sulfides, from pipelines. This is due to the comparatively much greater weight and tighter fit for the ILI vehicles.

6.1.5 Petal Pigs for Scraping and Cleaning Pipelines

Petal pigs are a specialty cleaning pig designed for cleaning multi-diameter pipelines. The name comes from the unique design of the scraper disks, which maintain a tight seal with the pipe wall, but are in the form of overlapping flower petals that can bend, enabling passage through the smaller diameter segments. Figure 34 illustrates petal pigs used to clean multi-diameter pipelines. The photo on the left side of Figure 34 is a clean pig, ready for a run through a pipeline, whereas the photo on the right side of Figure 34 illustrates a petal pig being retrieved from a pig receiver.

Petal pigs are uniquely suited for being able to transit through pipeline segments having different internal diameters, as well as removing paraffin deposition in crude oil pipelines. Figure 34 illustrates petal pigs, which were used to remove paraffin deposits from a pipeline having segments with different internal diameters. The hard plastic disks flex and scrape across the interior surface of the pipeline, displacing any deposits on the interior surface of the pipeline. As such, petal pigs could be used for displacing and removing black powder depositions. However, unlike cleaning pigs, which have spring-loaded brushes, such as illustrated in Figure 33 or those presented in the next section, scraper disk pigs cannot remove deposits from within pits extending into the base metal of the pipe from the internal surface of that pipeline. Instead, scraper disc petal pigs would push or smear black powder or corrosion products into any preexisting pit.

Figure 34 Petal Pigs for Cleaning Multi-Diameter Pipelines

6.1.6 Specialty Pigs Designed for Removing Deposits from Existing Pits

Figures 35 and 36 illustrate a specialty cleaning tool, which has been designed to remove debris from pre-existing pits or crevices along the interior surfaces of pipelines. The tiny fingers, which collectively appear similar to a rake, are spring loaded and ride along the interior surface of the pipe wall to be able to break up deposition in any pits.

Figure 35 Specialty Pigs to Loosen and Remove Deposits from Pits
Photograph courtesy of TD Williamson

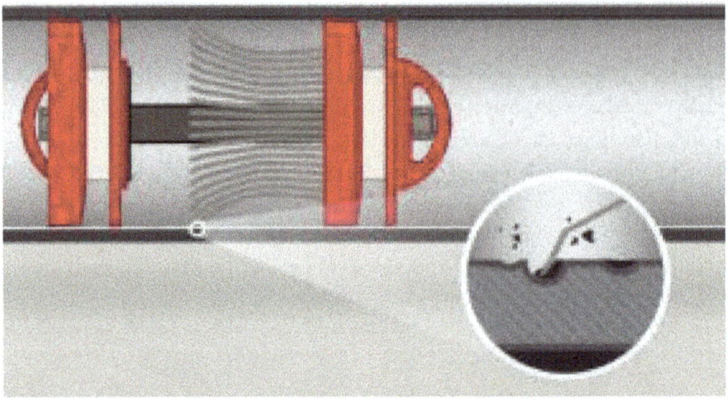

Figure 36 Cleaning Pipeline Pits Using Spring Loaded Fingers to Scrape Deposits
Although only one spring loaded finger is depicted, multiple fingers cover all orientations.
Photograph courtesy of TD Williamson

The value of these specialty pigs is their ability to enhance results from in-line inspection runs, which follow pipeline cleaning operations. If a pipeline has internal pits and black powder, maintenance pigging runs conducted with foam, petal, or scraper disk pigs would likely have smeared that black powder into those pre-existing pits, filling and packing them. However, the specialty pigs, such as illustrated in Figure 36, have the ability to scrape away at least some of the material packing the interior of any of those pits, thereby making the pits easier to detect by the follow-up inline inspection vehicles.

- Iron sulfide is magnetic, and as such, the depositions may mask the pits from easy detection when in-line inspection (ILI) vehicles are using magnetic flux leakage (MFL) technology to inspect the pipelines.
- Specialty pigs can help clean the pipelines, making it easier to detect pits.

6.2 Maintenance Pigging Programs to Keep Line Clean

The frequency of maintenance pigging runs needs to be determined on a case-by-case basis for each individual pipeline, based on the volume and characteristics of the material removed and ability to handle that volume of solids or liquids. There should also be a minimum frequency of those pig runs, such as quarterly or annually, because results from such runs serve as an indicator for potentially changing conditions. Those changing conditions may or may not warrant the initiation of remedial measures.

Figure 37 Black Powder Sludge Removed at Pig Receiver

6.2.1 Adjust Pigging Frequency to Keep Volumes Removed Manageable

The volume of material removed during any individual maintenance pigging cleaning run must be controlled, such that the valve controlling the flow, either into the pig receiver or into the bypass loop, can be closed to isolate the pig receiver. That means the volume of the material removed, plus the volume occupied by the cleaning pig, cannot be greater than the volume of the pig receiver, as measured from the isolation valve. Otherwise, the valve could not be closed, and there would be safety concerns. The volume of material removed during any individual cleaning run can be controlled by (a) controlling the potential formation of corrosion products, such as through the use of corrosion inhibitors or hydrogen sulfide (H_2S) scavengers, (b) selecting maintenance cleaning pigs, based on their aggressiveness or ability to remove black powder or corrosion products, or (c) adjusting the frequency of maintenance cleaning pig runs. This section will address the third option–adjusting the frequency of maintenance pigging runs, based on historical records, such that the volume of materials removed at manageable levels is kept manageable.

Consider one pipeline in West Virginia, which transports natural gas and natural gas liquids (NGL) to a processing facility. That one particular pipeline had to be pigged three times a day to prevent liquids from accumulating at low spots and choking or restricting flow. Such a frequency is very unusual–quarterly or semi-annually is far more common. However, it illustrates that the pigging frequency should be based on knowledge of the operating conditions for any individual pipeline and doing what is necessary for continuing operations: flow assurance. The goal is to maintain operations, while keeping the volume of solids or liquids removed from any individual pipeline at manageable levels, and without disrupting production.

If the material removed is pyrophoric iron sulfide, the recommended frequency for maintenance pigging may also need to be adjusted. This is based in part on the total volumes of water required to flush any black powder off the pig, once the pig is received. Figure 2 illustrated such an operation to flush and clean the pig receiver and cleaning pig within a lined pit, once that pig was removed from the receiver. The black powder removed from the pipeline and the maintenance pig itself needs to be kept wet until it is turned over to a hazardous waste facility. That facility also has to have the capability to process the volume of liquids used in such a cleaning/flushing operation.

The frequency might also need to be altered if the black powder or other debris removed contains naturally occurring radioactive material (NORM), because there would be different handling requirements.

Pigging Frequency

Adjust the frequency of maintenance pigging to keep the volumes of solids or liquids removed during any individual pig run at manageable levels.

6.3 Example: Pigging Samples Used to Assess Effectiveness of Remediation

The following is an example of using sample analysis to assess the effectiveness of a chemical remediation program. The example is significant for several reasons. It illustrates the use of sample analysis for monitoring conditions in the pipeline and the effectiveness of chemical treatments. This example deals directly with controlling populations of sulfate reducing bacteria (SRBs) and their release of H_2S. That, in turn, directly relates to the formation of black powder, which will be discussed here.

The particular pipeline highlighted in this example was breached when a lateral was ripped from the subsea trunkline during the passage of Hurricane Ike.[52–53] The pipeline filled with untreated, oxygenated seawater, plus sand, because the trunkline was on the bottom of the Gulf of Mexico. The seawater contained both sulfates and SRBs, and the interior of the breached pipeline provided the perfect environment for the population of the SRBs to flourish, as well as contain the H_2S these microorganisms generated. The seawater also contains dissolved oxygen up to approximately 10 ppm, depending on temperatures. Consequently, all the ingredients were in place to form both iron oxides and iron sulfides–the primary components of black powder. This example is even more significant in that tetrakis(hydroxymethyl)phosphonium sulfate (THPS), a biocide and a chelant, was used to reestablish control over the microbiological populations, as well as to chelate and remove any of the black powder that may have formed. Chelants are discussed in Section 13.3, and the chelant properties of THPS are discussed in Section 14.1.

Please refer to the original papers[52–53] for discussions of the configuration of the trunkline and lateral, and how the pipelines were reconnected. Once the trunkline was repaired, the first step was to remove the bulk of the seawater from the trunkline by using onshore sources of natural gas and pushing a foam pig from the onshore facility to the offshore production platform, where the pipeline flow normally starts. This reverse direction pigging was necessary, in that the onshore facility did not have the capacity to handle the volume of seawater, which had to be removed from the trunkline. The initial dehydration operation was to displace the bulk of the seawater to the offshore platform and into an appropriately sized barge, such that that seawater would be captured and later transferred to an appropriate disposal site. Subsequent pipeline dehydration and chemical treatment runs were then conducted from the offshore platform to the onshore processing facility, once offshore production resumed.

From a safety and health perspective, one of the greatest concerns associated with the dehydration and remediation of the pipeline was the concern related to potential concentration of H_2S dissolved in the seawater being removed, or in the gaseous phase. This was based on knowledge that the seawater contained both sulfates and SRBs, and that any H_2S generated by those microorganisms would be contained in the breached pipeline. Hence, it was a top safety and health priority to measure the concentration of the H_2S in the seawater and gas being removed from the pipeline to properly assess that threat. This would also provide an insight into the potential formation of any iron sulfide in that breached subsea pipeline.

The concentrations of H_2S were measured using techniques, such as discussed in Sections 5.1.2 and 5.1.3, using ampoules or stain tubes designed for measuring much higher concentrations than encountered in normal operations. The measurements for the concentrations of H_2S were initially collected offshore during the initial dehydration run, which flowed to the offshore platform and awaiting barges. Later, the measurements were conducted onshore, once the offshore production resumed.

Prior to the breach, there was at most, only a trace of H_2S in the natural gas from the offshore gathering system. The natural gas from an onshore transmission pipeline, which was used for displacing the bulk of the seawater from the trunkline onto the awaiting barge, had at most, a trace of H_2S. When samples of the seawater were collected just ahead of the arrival of the initial dehydration pig, the concentration of dissolved H_2S spiked to 90 ppm. Later, when the associated lateral was dewatered, the initial reading of dissolved H_2S spiked to 10,000 ppm, but subsequent measurements dropped to between 500–1200 ppm H_2S. These measurements clearly demonstrated the microbiological activity of the SRBs, and those concentrations of H_2S can certainly generate significant volumes of iron sulfide corrosion products. (Also note the H_2S residuals within the lateral were higher than those of the trunkline due to 70 additional days prior to dewatering the lateral, which was in warmer, shallow water.) Those concentrations of H_2S can certainly generate significant volumes of iron sulfide debris.

The chemical treatment program consisted of applying (a) scavengers, (b) THPS, and (c) corrosion inhibitors, plus an accelerated maintenance pigging program. Samples were collected at the onshore facilities to characterize the solids and liquids removed, coinciding with the arrival of each cleaning pig. Thus, trends could be established, based on results from the analyses.

The priority for the initial chemical treatment was to quickly bring the microbiological population of SRBs into check, and then to remove corrosion products and sand. Fluid samples were collected and analyzed for months; each sampling coinciding with the arrival of each maintenance pig at the onshore facility. The analytical results demonstrated the overall effectiveness of the maintenance pigging and chemical treatment program. Figure 38 illustrates the treatment program and monitoring results.

Based on sampling results, iron sulfide was not seen as a concern. The original explanation posed was that the oxygen in the seawater, which entered the pipeline following the breach, may have reacted with the iron in the pipe wall, forming an iron oxide scale that helped protect the base metal from the H_2S the SRBs generated. However, a better explanation may be that after the offshore production resumed, the injection of H_2S scavengers, coupled with large concentrations of THPS to re-establish control over the microbiological populations, and the application of corrosion inhibitors was the right choice. The application of THPS required large batch treatments, which provided the concentrations for chelant functionality, and hence, THPS was able to remove the iron oxides and sulfides.

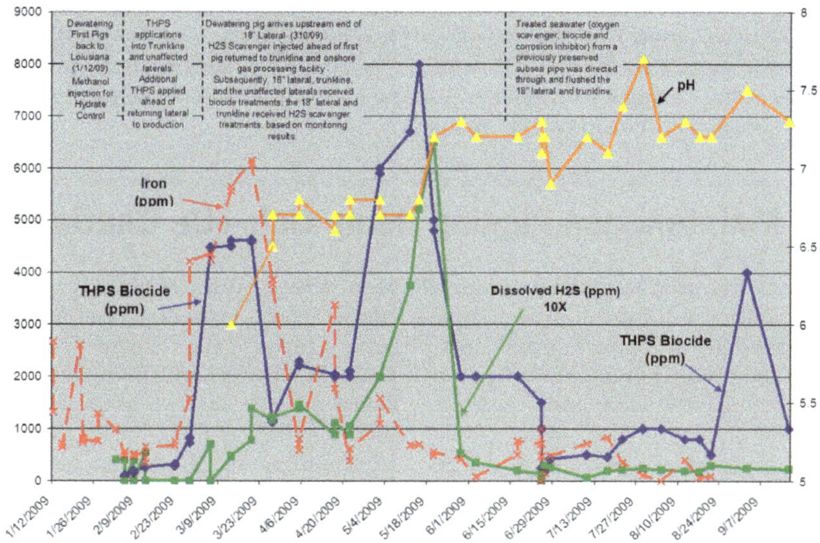

Figure 38 THPS Used to Remediate Pipeline Having SRBs and 90+ ppm H_2S[52]
The left *y*-axis is in parts per million, while the *x*-axis provides a timeline.

6.3.1 Goal: Remove Black Powder by Pigging and Chemical Treatments

Chemical treatments can be used in conjunction with maintenance pigging to remove black powder from a pipeline. The historical barometer or measure used to quantify the benefits of the chemical treatments was a measure of the iron counts in the fluids captured at the pig receiver, which is a NACE standard.[54] However, more recently, the measurement for the concentration of iron is just one of many parameters used when assessing the effectiveness of chemical treatments. Generally, analyses of samples collected from several successive maintenance pig runs are necessary to document the effectiveness of chemical treatment programs.

When chemically cleaned using a THPS solution, the initial iron counts could be very high, such as 30,000 ppm or approximately 3%. However, through several cleaning runs, the iron counts can be drastically reduced to around 250 ppm–a very significant two orders of magnitude reduction in the parts per million of iron.

Another challenge is to determine when the pipeline should be cleaned again. Unfortunately, this needs to be established on a case-by-case basis, because many parameters come into play when determining the rate black powder would otherwise accumulate in the pipeline. Compare the concentration of H_2S from the upstream source(s) versus the downstream concentrations of H_2S, and use the difference in concentrations, coupled with the production rates and time between pigging runs, to estimate the potential volume of iron sulfide, which could theoretically be removed from the pipeline. Then compare the estimate to the actual volumes removed during the maintenance pig run. Such data will help determine the necessary frequency for chemical treatments and maintenance pigging runs, so that black powder can be kept in check.

6.4 Maintenance Pigging Using Gels—Be Cautious

Specialty gel products have occasionally been developed for cleaning pipelines. The applications of such products typically fall into one of two categories (a) forming a solid gel in a pipeline, and then pushing it through the pipeline to remove unwanted product or depositions, or (b) injecting a sticky gel, which would traverse the pipeline, picking up and transporting debris from the interior of the pipeline.

Based on product descriptions, the gel products presently offered are essentially for forming solid gel pigs in otherwise non-piggable pipelines, and using those gel pigs to displace liquids, but not solids, out of those pipelines, which have been taken out of service.

One service provider cited using a cross linked, food grade polymer, which was not hydrocarbon soluble, and reported achieving a 58% increase in flow through a 6-in (15.2 cm) flow line from 190–300 thousand cubic feet per day (MCFD) (5.38–8.50 million cubic liters per day [MML/d]), after 13 barrels (bbl) (546 gal = 2067 L) of water and 3 bbl (126 gal = 477 L) of condensate were removed. The service provider noted the flow was directed into a large swab tank, and not into or through any downstream processing equipment. As such, they avoided a potential pitfall of small pieces of the gel breaking and plugging any orifices in downstream processing equipment.

Caution must be exercised regarding the use of loose, sticky gels, which theoretically would adhere to pipeline debris and transport that debris to a point downstream, where it could be collected and removed. Specifically, concerns were voiced since the 1990s that a small portion of the loose, sticky gel could become lodged within any sensors penetrating through the pipelines or process equipment, compromising the functionality of such production instrumentation. Based on those concerns, carefully assess systems before introducing any gel products to ensure any unintentional, deleterious consequences are not introduced.

Before leaving the topic of gel pigging, the results from one additional field study of a highly cross-linked gel pig should be cited.[55] This particular pig was formulated using food grade gelatin and a chemical that cross links the gelatin, making it more stout and able to pass through the pipeline without breaking up. Corrosion inhibitor was added to the gelatin before it gelled or solidified within a short bypass loop that served as a joint pig mold and launcher. The purpose of that study was to determine whether the corrosion inhibitor could be better applied to the pipe wall through an ablative process by the gel pig being pushed from behind and flexing, such that the gel pig rubbed against the pipe wall. The effect would be similar to that of an eraser being rubbed against the pipe wall.

The pipeline selected for this particular test operated at temperatures slightly above the melting point of the cross-linked gel. This particular pipeline also did not have a pig launcher nor receiver, which could have provided an option for maintenance pig to clean this pipeline. Accordingly, a plan was drafted to install a short line loop, which could be isolated and would serve as a mold for forming a cross-linked gelatin pig. The concept was that such a cross-linked pig would be sufficiently stout to transit the pipeline, displacing any deposits, while at the same time rubbing against the pipe wall, and applying the corrosion inhibitor through ablation. The application would be unique in that no receiver would be needed, because the length of the pipeline was sufficiently short that the gelatin pig would arrive and enter the downstream slug catcher vessel before melting. Also, the residency time within the slug catcher vessel was sufficiently long as to guarantee such gel pigs would quickly melt or dissolve due to the operating temperature.

The performance measure used to document the effectiveness for this particular methodology for applying corrosion inhibitor to pipe walls was based on the measured corrosion rates. Success would mean a reduction in the measured corrosion rates. However, there were no measured reductions in the corrosion rates, based on field tests. Consequently, this concept for using solid gel pigs to apply corrosion inhibitors was abandoned.

7 In-Line Inspections and How They May be Affected by Black Powder

In-line inspection vehicles (ILI), commonly known as "smart pigs," are highly developed tools used to inspect pipelines and identify a precise location for any anomalies in the pipeline between the pig launcher and receiver. Unlike radiography, or ultrasonic inspections, which are used to inspect localized segments of a pipeline or process equipment, ILI vehicles continuously collect data around the entire circumference of a pipeline, and along the entire length of the pipeline. Each ILI vehicle has numerous sensors around the circumference of the pigs. There is an overlap in the coverage for each sensor, such that there are no blind spots, even if individual sensors are damaged or fail during an inspection run. The inspection data is collected continuously as the ILI vehicle passes through the length of the pipeline. The continuous data collection makes for massive data files, and the analysis of that data is typically conducted by the ILI service providers in their main offices.

The service companies and pipeline operators commonly compare the most recent inspection results to the results from the previous inspection run through the same pipeline. This is to determine whether an area of interest may be enlarging over time, which could affect the integrity of the pipeline. There is a temptation to estimate the corrosion rate for a small area by comparing the most recent results to previous readings. However, the ILI service providers state their results are at best accurate to 10% of wall thickness. That can be pretty coarse, compared to weight loss coupon results, iron counts, or even localized radiographic (RT) or ultrasonic (UT) inspections.[46, 54] For that reason, ILI results are commonly used to assess the overall integrity of the pipeline, as well as identifying specific or critical locations, which should be monitored or inspected more frequently.

The discussions below will note possible limitations in the results from ILI runs whenever the pipeline anomalies, or pits, are packed with dense, magnetic black powder. See Section 6.1.6 for a discussion of specialized cleaning tools for removing deposits in pits.

7.1 Overview of MFL and UT ILI Vehicles

There are two main technologies used for conducting ILIs. The first is based on magnetic flux leakage (MFL), and the second is based on ultrasonic (UT) technology.

MFL inspection vehicles have numerous magnets around the entire circumference of the pig. The fields can be oriented along the axial direction of the pipe, or at a different orientation, depending on whether the inspection is to identify corrosion pits, or perhaps cracks having unique orientations. The residual magnetic fields are detected, using sensors that are typically located between the magnetic poles or farther back along the axis of the ILI vehicle. A strong magnetic field is established in the pipe wall as the pig transits through the pipeline. The key term is "saturated," meaning the magnetic field in the pipe wall has the maximum strength. When inspection vessels pass or travel through the pipeline, they establish strong, maximum strength, magnetic fields. The sensors detect any disruptions to the magnetic field as the pig passes through the pipeline.

UT inspection vehicles are similar in that they have ultrasonic transponders around the entire circumference of the pig, much like the arrangement of magnets and magnetic field sensors. Both collect signals essentially continuously from the time of launching until arriving at the receiver.

Both types of vehicles will also have odometer wheels to measure the passage through the pipeline. The sensors will detect each girth weld, enabling positions along the pipeline to be determined accurately, using the known position of the upstream and downstream girth, and matching the data to those known locations.

7.1.1 Magnetic Flux Leakage (MFL) Technology

Figure 39 is a simplistic drawing to illustrate the passage of saturated magnetic fields through the pipe wall.

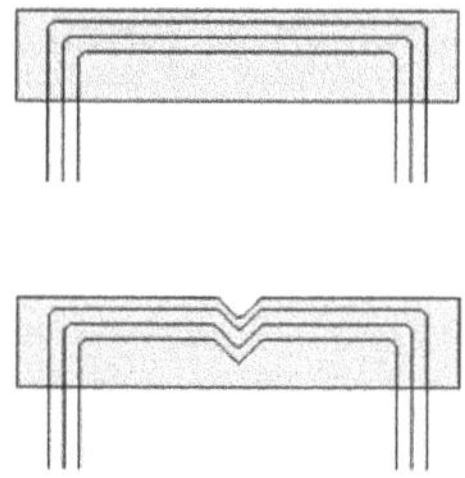

Figure 39 Magnetic Flux Leakage Technique for ILI Inspections
The magnetic field saturates the base metal. Sensors located between the magnetic poles can detect variations of the magnetic field, which are used to determine the location and characterize any pipeline anomalies.

"Pipeline anomaly" is a term used to identify a suspected defect on a pipeline before it has been confirmed through independent inspections. MFL tools have the best chance for detecting pipeline anomalies when the direction of the magnetic field is perpendicular to the direction of the anomalies. Thus, for example, if circumferentially oriented cracks are suspected along the girth welds, the best chance to detect such cracking would be through orienting the magnetic field axially, along the length of a pipeline. Likewise, if axially oriented cracks are suspected, they would be best detected through magnetic fields that are circumferentially oriented, around the pipeline. This is called "transverse MFL." Still other MFL inspection vehicles have employed magnetic fields at a 45° angle relative to the axial direction. Although not optimized for either axial or circumferential cracks, it is a good practical and cost-effective compromise, which provides the ability to detect anomalies having varied or unknown orientations.

> Please consult with the in-line inspection service providers to select the best tool for detecting the type of anomalies suspected for any particular pipeline.

7.1.2 Ultrasonic (UT) Technology

Consider the functionality of UT ILI vehicles. The UT ILIs transmit a sound signal into the base metal and then listen for the return signal. Note that a liquid couplant is required for passing the UT signal from the transducer to the base metal, as well as passing the reflected signal back to the transducer. Sound travels very quickly through steel, approximately 20,000 ft/s (6096 m/s). The wall thicknesses are typically less than an inch (12 in = 1 ft, 1 in = 2.54 cm). The time for an ultrasonic signal to (a) pass through a thin liquid couplant located between the UT transducer and pipe wall, (b) through the base metal of the pipe, (c) reflect off the back surface of the pipe wall or any anomaly, and (d) then pass back through the couplant to the transducer, would be in the range of milliseconds. This is illustrated in Figure 40.

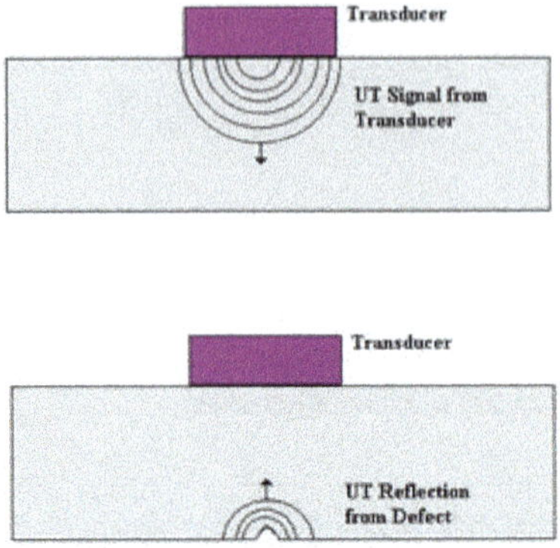

Figure 40 UT Technique for In-Line Inspection of Pipeline
The UT signal passes through a thin liquid couplant (gel) between the transducer and pipe wall.

7.1.3 UT Technology—Overlapping Readings Ensures Complete Inspection

The text above describes the transmission of the UT signal, and then listening for the return signal. However, the ILI vehicles are continuously traveling through the pipeline. Hence, the natural question is whether significant data could be missed when the UT transducers are transmitting their signals, as opposed to listening. The quick answer is "no," and here is the explanation.

Consider the speed at which ILI pigs travel while conducting their inspection runs. The optimum speed for ILIs is typically 5–7 miles per hour, or approximately 7.3–10.3 ft/s (2.23–3.14 m/s). The time required for the UT signal to travel to/through the pipe wall and then return is in the range of milliseconds. The distance an inspection vehicle would travel in that time is in the range of a hundredth of a foot (3 mm), or approximately half the diameter of a ¼-in (6.35 mm) transducer. Accordingly, each of the signals and return signals overlap, thus effectively providing continuous readings along the path for each transducer as the ILI vehicle travels through the pipeline from the launcher to receiver. Since the UT transducers are placed throughout the circumference of the ILI vehicles, and each is continuously collecting data, such vehicles provide a comprehensive survey along the length and circumference of the pipeline from the launcher to the receiver.

7.1.4 Electro-Magnetic Acoustic Transducers (EMAT) Technology

Electro-Magnetic Acoustic Transducers (EMAT) is an ultrasonic inspection technique, which is unique in that it generates the sound in the pipeline or tubing being inspected, as opposed to using acoustic transducers. The ultrasonic waves are generated in the test material by electromagnetic induction, using two magnetic fields that interact and generate a Lorentz force. The force is transferred to the crystal lattice of the material being inspected. Those forces, in the form of waves in the crystal lattice and the applied magnetic field, induce currents in receiving EMAT coils.

The technique is adept at detecting cracks on pipelines or pressure vessels. Literature citations were not found to address whether there were any deleterious effects in the ability to detect cracking if iron sulfide deposits filled those cracks.

7.2 Illustration of Typical ILI Results

The ILI service providers typically offer options for displaying the results from the inspection runs. Figure 41 illustrates results for a localized segment having several pipeline appurtenances (weld-o-lets), along with a girth weld. Whenever a pipeline has a history for having black powder, be aware of the possibility that black powder may be smeared into any existing pits, making those pits more difficult to detect. (Black powder is typically evidenced by the material removed during maintenance pigging cleaning runs that precede inspection runs, samples removed from sample ports, or perhaps based on measurable volumes of black powder removed during previous inspection runs.)

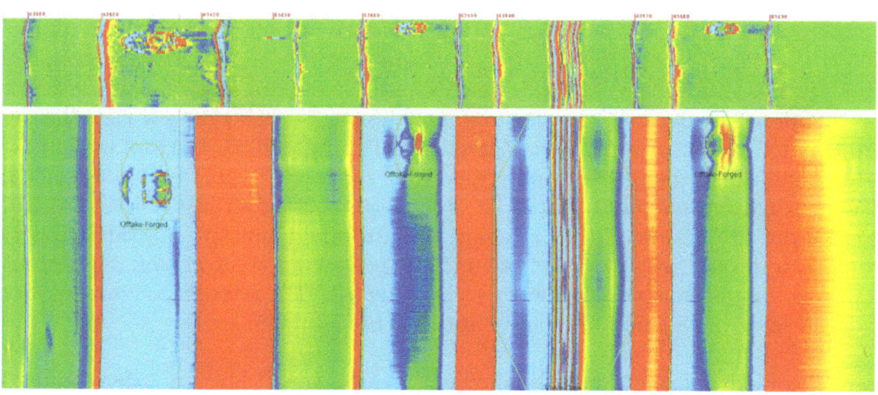

Figure 41 Illustration of One Type of Display Depicting Results from ILI Runs
This particular segment illustrates several forged offtakes.

The ILI run will provide the specific axial location and o'clock orientation for the pipeline anomalies or indications, which were detected. The data will also provide the depth, length, and width for the most significant anomalies. The standard practice is to review the results from the run, and select a few of those anomalies or indications for further investigation. These are typically some of the most significant or deepest of the anomalies, and they are investigated to confirm the inspection results. This would typically require an excavation at the location of the reported anomaly. As appropriate, the results from the original inspection run could then be adjusted, based on results from comparing the original field inspection results to actual measures of those specific anomalies following pipeline excavations.

There may be limitations regarding the accuracy or validity of any inspection run. It was noted above that any existing black powder could have been smeared into preexisting pits on the interior of the pipeline, making them more difficult to detect. However, black powder is not the only potential limitation. For example, if inspecting a pipeline, which has a history of paraffin deposits, be aware that a fine layer of paraffin deposition can continue to build on the pipeline from the time of the cleaning run to the inspection run. Such paraffin deposits can compromise inspection runs using UT ILI vehicles because the UT signals are attenuated. However, MFL ILI vehicles are less affected by the presence of paraffin, since the magnetic fields can pass through thin layers of paraffin. Pipeline cleanliness is essential for obtaining good, valid inspection results.

7.3 Limitations on Quality of ILI Results When Black Powder is Present

Consider the affects that black powder may have on the quality of results from an inspection run. Black powder is typically a mixture of iron oxide and iron sulfide, both of which are magnetic materials. Assuming a pit is present, that pit probably became filled with magnetic black powder, and the packing would become denser as additional black powder is smeared into the pit through multiple cleaning pig runs. Hence, for example, the magnetic field from an MFL ILI tool could pass through those magnetic depositions in a pit and appear like unaffected base metal. For that reason, pipelines should be cleaned just before an inspection run, and one of the last pigs used in the cleaning process should have spring loaded fingers or similar devices, which can cut into and gouge the comparatively soft depositions packing the pits making them more easily detected. The referenced cleaning tool was presented in Section 6.1.6 and illustrated in Figure 35.

The complementary question is whether the presence of black powder could affect the ability for an inspection tool to detect pipeline cracks. Note that hoop stresses are greatest on the exterior surface of a pipe. That is where cracks are found–not on the interior surface. Thus, black powder within the pipelines could not pack exterior cracks, which are typically found on uncoated sections of steel pipelines or under disbanded coating. Accordingly, any accumulation of black powder within a pipeline could only attenuate the magnetic fields or ultrasonic signals, which ILI vehicles use to detect and quantify pipeline anomalies. Again, use maintenance pigs to clean pipelines prior to inspection runs.

7.4 Considerations if Using Water as Couplant: Do Not Leave Water in Line

Although UT ILI vehicles provide good, valid inspection results, they are not always selected, because UT transducers require a liquid couplant between the transducer and the metal being inspected. This would not be a problem for pipelines transporting certain hydrocarbon liquids or perhaps water in water injection systems. However, there would be a reluctance to fill a dry gas transmission pipeline with water just to conduct an inspection run. Consider the challenge in completely removing that water before operations resume. For those cases, the use of MFL-ILI vehicles makes more sense, as the magnetic field does not require any couplant.

Nevertheless, there are some special case inspections, which are best when conducted using ultrasonic technology, such as looking for cracks having a particular orientation. UT inspections require a couplant, and the most practical options for providing the required couplant are to either fill the entire line with water (or another acceptable liquid), or to establish a pigging train. The pigging train would encapsulate the water couplant around an UT-ILI vehicle, using leading and trailing foam pigs that have enhanced capabilities to form a seal with the pipe wall to retain sufficient couplant for the inspection run. Such a pigging train is illustrated in Figure 42 below.

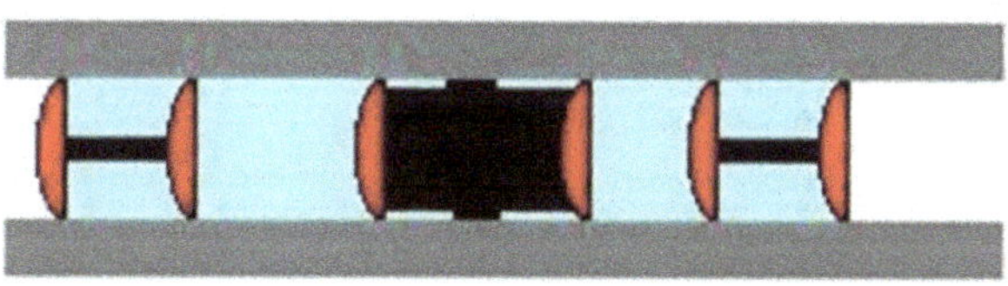

Figure 42 UT Pig Train to Provide Couplant for UT ILI

The establishment of such a train is more complex than launching a single pig, and would require coordination between several operations personnel. For example, it may be necessary to launch the front sealing pig, shut down the pipeline's flow, fill the line between the first pig and the launcher with the water couplant, launch the UT-ILI inspection vehicle, shut down operations, fill the line behind the UT-ILI vehicle with additional liquid couplant, and then launch the rear sealing pig. When receiving the pigs, it may also be necessary to suspend operations when receiving each pig, unless the receiver has the length to accommodate all pigs. Alternatively, the receiver can be modified by using pig trap extenders, which increase the functional length of the pig receiver.

During such a run, expect some loss of couplant around the seals at the front and rear of the pig train as the three pigs transit through the pipeline. Adjust the volume of the couplant to account for such loss. The in-line inspection service companies should be able to provide some guidance, based on their collective experience in conducting numerous inspection runs.

Following completion of any inspection run, which uses water as the couplant, efforts should be initiated to remove any remaining water from that line, particularly if there may be an extended period of time (e.g., greater than 28–30 days) before operations resume.

Chapter 8 provides an example that illustrates the unintended consequences from a small volume, or "heal," of water that remained within a pipeline for an extended period of time, following a hydrostatic pressure test. The concern relates to the possible consequences if those test waters contained SRB, which establish colonies and generate the highly acidic H_2S that could dissolve base metal and form a hole that penetrates the pipe wall. The same concern relates to pipelines, which were inspected using UT tools that required the introduction of water. Operations must resume shortly after completion of the inspection runs, or that water must be removed. The rule of thumb, as presented in Chapter 8, is to treat the line if the downtime prior to the resumption of operations exceeds 28–30 days.

8

Hydrostatic Pressure Tests to Document Integrity—Beware Any Remaining Water

The three acceptable methodologies for demonstrating the integrity of pipelines are (a) hydrostatic pressure tests, (b) in-line inspections, and (c) the NACE Direct Assessment methodologies. The pipeline owner/operator selects the methodology or methodologies for use on individual pipelines.

Hydrostatic pressure tests are typically conducted immediately after completing construction of a pipeline or processing equipment. However, hydrostatic pressure tests are also conducted in operating pipelines, such as within crude oil or natural gas gathering systems, where it may be very difficult or expensive to conduct in-line inspections or the direct assessment methodologies. Although hydrostatic pressure tests can be conducted on existing, operational, gas transmission pipelines, it is not often undertaken due to the challenges in removing absolutely all the water following the test.

Personnel conducting the hydrostatic pressure tests typically use test water obtained from local streams and rivers, due to the expense and/or limited availability of large volumes of municipal water. Also, there may be significant costs associated with transporting high quality potable water to remote locations for conducting hydrostatic pressure tests on those pipeline segments. The plan is generally to return that water to the original source, following completion of the test. It should be fairly simple to obtain the discharge permits for testing new pipelines, which have never transported hydrocarbon products. However, a more cumbersome process would be expected when discharging water following hydrostatic pressure tests of operational pipelines that had been transporting fluids containing hydrocarbons. The challenge is water quality—both for the source and for discharge.

When new pipelines were recently being constructed in Alberta, Canada or Pennsylvania, U.S.A., local water sources, including lakes and ponds were tapped. Serial dilution tests were conducted to quantify the population of both sulfate reducing bacteria (SRB) and acid producing bacteria (APB). The populations of both SRBs and APBs were reported as high as 10^5, or 100,000 microorganisms per milliliter. However, test analyses of the source water were not available before the hydrostatic pressure tests were conducted. (The serial dilution tests took 28 and 14 days to complete, following sample collection.) Consequently, the pressure tests were completed and the water was discharged (returned to the original source) long before results from the serial dilution studies were available. As a result of using poor quality water, the chances for microbiologically influenced corrosion (MIC) and the formation of black powder increased due to the microbiological population in the test water that remained within the pipeline, even after pipeline dehydration operations.

If those pipelines went into service shortly after completion of those tests, i.e., within 28–30 days, there would generally be no concerns related to the threats to pipeline integrity posed by SRBs or APBs. However, if there was an extended period of time before operations began, which is typically greater than 28–30 days, then remedial measures should have been implemented because the threat posed by the residual test water and the microbiological population within that water.

Figures 43 and 44 are photographs showing the exterior and the interior of a through-wall pit, which occurred in a new pipeline that failed before it went into service.[56] Although the pipe had successfully passed the hydrostatic pressure test, it was almost a year from the time of the hydrostatic pressure test until the time when construction of the associated facility was completed and production was scheduled to start. This particular failure was attributed to microbiologically influenced corrosion (MIC) from SRBs that entered the pipeline in the water used for the hydrostatic pressure test. The pits provide evidence that the pipe had not been completely dehydrated. Localized colonies of SRB flourished, and their waste product, H_2S, is highly acidic, and generated the round, smooth, pit-within-pit appearance.

Figure 43 Pinhole Leak Within a New Pipeline Attributed to MIC
There was a 1-year layup following the hydrostatic pressure test before operations began.

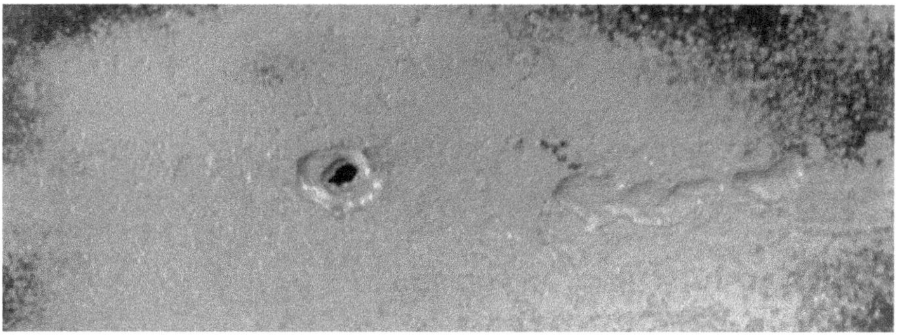

Figure 44 Interior View of Pinhole Leak, Evidence of MIC
The interior surface was black before cleaning, suggesting possible black powder.

The NACE paper (Reference 56) provides guidelines and options for treatments, which can be used to preserve the pipeline for extended periods of time, while at the same time being cost effective. The assumption is that personnel, who conduct hydrostatic pressure tests following fabrication of pipelines, will use locally available water for those pressure tests.

Unfortunately, the practical reality is that poor quality water is often used at locations away from potable water sources, such as municipal systems. Although the test water used may be filtered, water from ponds or streams often contain high populations of microbes, such as SRBs and APBs, which will not be removed by filtration, and will pose a threat to the integrity of pipeline networks if left within the pipeline for extended periods of time (generally taken as greater than 28–30 days). The "trick" for preserving pipeline integrity is to remove as much of the test water as quickly as practical after completing the test, and then chemically treat the remaining heal or small volume of test water if pipeline operations will not begin within 28–30 days. Here is a strategy that has been employed:

8.1 Strategy for Treating Pipeline if Extended Time Following Hydrostatic Pressure Test

Following successful completion of hydrostatic pressure tests, that water would be removed from the test segment, using foam pigs to dehydrate the line. Forced air may also dry the pipeline to the point of meeting dew point or moisture requirements. However, as seen in the example above, a significant volume of water may still remain within the pipeline, such as at low points, as well as at each girth weld along the length of the pipeline.

Although general estimates predict 95% of the volume of water that had been within the pipeline has been removed, a 10% volume is used for calculations. The concept of the preservative treatment is to take a special batch of potable water and add a combination biocide, corrosion inhibitor, and H_2S scavenger product. The volume of the biocide component would be 500 ppm or greater, based on the volume of the special batch of potable water and the assumed 10% volume of the hydrostatic test water still remaining within the pipeline segment being tested. The treatment would be applied to the pipeline in the final pig run, wherein the batch of treating chemical would mix with any residual hydrostatic pressure test water, and any leftover volume from the batch would be captured at the pig receiver for appropriate disposal.

Such a treatment methodology is designed to preserve the integrity of pipelines whenever there is an expectation for an extended dwell time before pipeline operations would begin. Again, it is targeted to control SRB and the potential generation of H_2S, which would result in forming iron sulfides, the distinctive component of black powder.

9 Overview of Corrosion Monitoring Including When Black Powder May be Present

Virtually any pipeline or processing system is susceptible to both external and internal corrosion. The real question is "How great is the risk or potential for internal and external corrosion?" Most readers are aware that the potential for external corrosion can be monitored by regular potential measurements at the test stations, rectifier readings to ensure the proper output of cathodic protection systems, and periodic pipeline surveys, such as close interval surveys to measure the potentials along the pipeline. In an analogous way, internal corrosion can also be monitored. It just requires a different set of tools. This chapter is an overview of the tools used for monitoring internal corrosion, and also identifies the methodologies that are most reliable, even when black powder is present.

Internal corrosion is typically monitored, using corrosion coupons or electronic probes that have been installed into the pipeline. Another tool for monitoring and assessing potential internal corrosion is through chemical analysis of samples removed from the interiors of pipelines or process equipment. The chemical analyses are results obtained from analyzing any solids or liquids removed through maintenance pigging programs.

By definition, corrosion monitoring is the systematic collection and analysis of data to assess the potential threat for corrosion. Corrosion coupons and electronic probes provide the most cost effective tool for assessing the potential threat for internal corrosions. The focus is to look for subtle changes, which may indicate the onset of corrosive conditions, such that remedial measures can be implemented to reduce those threats and maintain system integrity.

This chapter will start with an overview of the four components of a corrosion cell. The intent is to provide the reader with a straightforward explanation for why cathodic protection on the exterior of pipelines provides no benefits for the interior of the pipeline. In addition, it explains why both external corrosion and internal corrosion programs are needed–a question commonly asked by management. Note also that all four components of a corrosion cell are needed in order for there to be active corrosion.

Next, corrosion monitoring itself will be discussed, including how internal corrosion rates are calculated, and what the acceptable general and maximum pitting corrosion rates are, per a NACE recommended practice[54] that represents the industry consensus. The proper location for installation of monitoring coupons or electronic probes will also be presented.

This chapter concludes with a discussion of iron sulfide "bridging" of electronic probes, i.e., if and when highly conductive deposits form electrical "short circuits" either between probe elements or different portions of the same element. Those electrical shorts may compromise the validity of resistance measurements used in the determination of corrosion rates.

Also, be aware that if iron sulfide deposits form across isolation flanges on the interior of pipelines, those sections of pipe may not be electrically isolated. A rigorous maintenance pigging program may provide benefits in removing such electrically conductive depositions.

9.1 Corrosion Cell—Why Internal and External Corrosion are Independent

The reader must recognize that internal and external corrosion are independent of each other. It is a misconception that a good cathodic protection system will also protect the interior of a pipeline. It cannot. Internal and external corrosion cells are independent of each other. The discussion starts with a description of corrosion cells.

9.1.1 The Basics: Corrosion Cell—How Monitor Relates to Cell Components

Corrosion cells consist of four components: (a) an anode, (b) a cathode, (c) an electrical path between the anode and the cathode, and (d) an electrolyte, such as water. These components are shown in Figures 45 and 46. Corrosion mechanisms are kept in check by controlling these four components. If anyone of the four components is missing, there will not be any corrosion.[57]

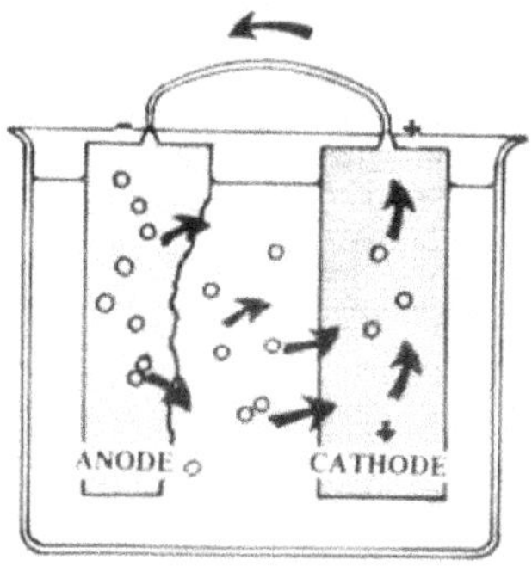

Figure 45: Four Elements of a Corrosion Cell[57]
The anode, cathode, electrical path between anode and cathode, and the electrolyte.

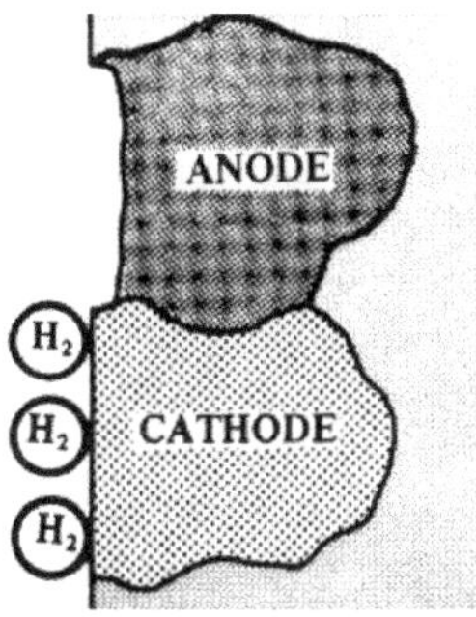

Figure 46: Anodic and Cathodic Areas on the Same Metal in Electrolyte[57]
The same four elements of a corrosion cell. Note the hydrogen bubbles in electrolyte. The anodic and cathodic areas are on the same metal and electrically connected.

There are numerous mitigation measures available. However, they all have the common theme of controlling at least one of the components of the corrosion cell to prevent corrosion. For example, coatings on the interior of tanks/vessels serve as a barrier for electrolytes contacting the base metals within those tanks or vessels. Corrosion inhibitors used within pipelines provide similar protective barriers, because they adhere to anodic or cathodic areas of the base metal to disrupt the corrosion cells. Biocides kill bacteria, and thus prevent SRB or APB bacteria colonies from secreting strongly acidic waste products that dissolve the base metal, underneath their colonies. Oxygen scavengers change the character of electrolytes, and dehydration helps minimize the presence of water in gas streams, thereby controlling the electrolyte, which is one of the components of the corrosion cell.

The effectiveness of any corrosion mitigation is typically measured through monitoring using corrosion coupons, which are small strips of metal inserted into the process stream for known periods of time. After a predetermined period of time, such as 90 days, the metal strips are removed, cleaned to remove any corrosion products, and then examined to determine the change weight over the period of exposure, which provides the general corrosion rate. Pitting corrosion rates are determined from the depth of the deepest pits formed over the period of exposure. Industry guidelines have been established to define generally acceptable corrosion rates. However, companies should establish their own maximum acceptable rates, based on operating conditions, such as pressure and temperature, the wall thickness necessary for pressure containment, and the remaining wall thickness. If measured corrosion rates exceed acceptable corrosion rates, remedial measures, such as corrosion inhibitor treatments, are warranted.

The coupons are the most cost effective of the measurement techniques, and have been the standard for decades. The metal strips are fabricated to match the metallurgy of the pipe wall, and provide a measure of the relative corrosion for that pipeline segment over the exposure period. Metal strips are electrically continuous, meaning that anodic and cathodic areas on the same strip of metal are electrically connected (touching each other). Thus, when wetted by an electrolyte, corrosion may occur. Corrosion monitoring via metal coupons is less susceptible to "bridging" or "electrical shorts" caused by the buildup of iron sulfide corrosion products, and as such would be the recommended monitoring tool in environments where black powder may be present.

Note, however, that other metrics may also serve as monitoring tools. For example, if a pipeline transmitting dry gas does not have any installed corrosion coupons, but does have moisture meters that *continuously* analyze each gas stream entering the transmission pipeline, then that line is indeed being monitored. There should not be any internal corrosion within the pipeline segment as long as there is no free, or condensed, liquid water to provide the electrolyte for the corrosion cell. Furthermore, any excursions in the moisture content should lead to remedial measures.

This chapter will primarily focus on monitoring for internal corrosion using corrosion coupons or electrical resistance corrosion probes, because those are the most commonly used and perhaps the most cost-effective tools for such monitoring and early detection of active corrosion. The goal is to detect and control corrosion mechanisms before there is significant damage to the base metals within the system.

9.1.2 Separate Corrosion Cells for Exterior and Interior Surface of Pipelines

There are separate corrosion cells for external corrosion and internal corrosion, in that there are different electrolytes on the exterior and interior of a buried pipeline. Part of controlling external corrosion is designing the system, based on the moisture content of the local soils. However, the soil moisture is entirely independent of any moisture within the pipeline. Thus, adjustments of cathodic protection systems will not have any effect on the internal corrosion mechanisms. Different corrosion control programs are required for the interior and exterior of the pipe. Anything done to control potential internal corrosion, such as the application of corrosion inhibitors, biocides, etc., will not have any effect on controlling external corrosion. Likewise, anything done to control external corrosion, such as the adjustment of the cathodic protection systems, would not affect internal corrosion.

9.2 Overview of Most Common Approach to Monitoring–Using Metal Coupons

The most common approach to monitoring potential internal corrosion is through placement of a small piece of metal coupon, such as depicted in Figure 47, within the pipeline for a specified period of time (such as 90 days), and then removing and cleaning that metal coupon to measure the weight–or more correctly the mass–loss over the exposure period. This mass loss over the test period is used to determine the corrosion rate, using the density or mass per unit volume of the metal used in fabricating the coupon and the pipeline. The depth of any pits can also be measured to determine the pitting rate, based on the total exposure period. Prior to cleaning a metal coupon removed from a system, the corrosion products or scale deposits, which built up on the metal coupons, can also be collected and assessed.

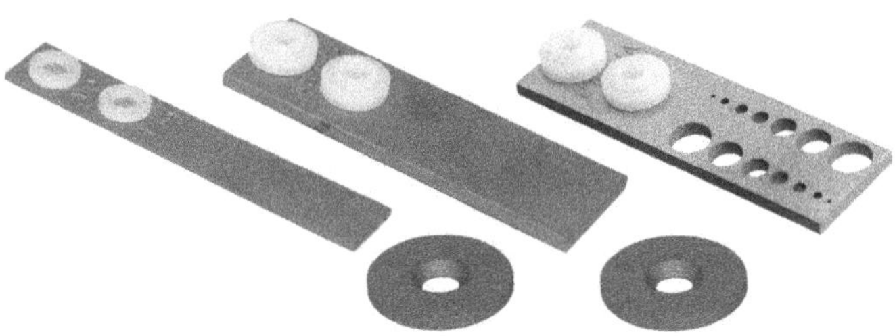

Figure 47 Manufacturers Offer a Variety of Metal Coupons for Monitoring
The coupon on the right is used to assess scale depositions.
Photograph courtesy of Rohrback Cosasco Systems

Metal coupons have a known geometry, whether it is in the form of a rectangular shape, a cylindrical shape, or appears like a round washer. The initial dimensions of the metal coupon will determine the total surface area exposed to the localized environment within the system being monitored. The density of the metal coupon, which is the mass divided by volume, is known. The particular type of metal coupon should match the metallurgy of the system being monitored, and be fabricated to have the same alloy or grade of steel, including heat treatments. This helps ensure the monitoring results are representative of any corrosion the pipe itself might experience.

Other factors that must be known for the corrosion rate calculations include the date or time the metal coupon was installed into the system, as well as the date or time that the metal coupon was removed from the system being monitored. The final piece of data needed for calculating the corrosion rate is the weight of the coupon after it has been removed from the system and cleaned to remove any deposition or corrosion products. There may also be benefits from documenting the appearance of the coupon as soon as the coupon is removed from service. This documentation includes the color and depth of any deposits or corrosion products once they are removed from the system and exposed to the atmosphere.

Prior to cleaning the metal coupon, it may be beneficial to conduct a field test to identify whether the corrosion product may be an oxide or a sulfide. Scrape a small sample of the corrosion products or deposition from the coupon and place it in a small glass bottle within a well-ventilated area. Then put a small drop of hydrochloric acid onto those corrosion products. The reaction can help determine whether those corrosion products are iron oxides or iron sulfides. Lead acetate paper is used to determine whether sulfides are present: the white lead acetate paper would turn brown in response to sulfides. If the sample fizzes and turns the color of a small strip of (prewetted) lead acetate paper, which is placed immediately below the bottle cap, and emits an odor of rotten eggs, then the corrosion product probably contains iron sulfide. However if fizzing, without any smell of rotten eggs or change in color for (prewetted) lead acetate paper, is observed, the corrosion product is most probably iron oxide, which can be confirmed by laboratory analysis. (Note that glass plates are sometimes used when bottles are not available. Although the fizzing from the reaction to the drop of hydrochloric acid can be observed, any hydrogen sulfide gas generated would not be contained, and could pose a health hazard. Best practices call for such tests to be conducted using bottles in well-ventilated areas.)

The analyses of those deposits or corrosion products can be confirmed by chemical analyses or perhaps XRF and XRD techniques.

- Corrosion coupons or probe elements should be fabricated from carbon steel having a similar chemistry and microstructure as the carbon steel of the pipeline.
- When inserted into a pipeline, the coupons or probe elements will react to corrosive conditions the same as the pipeline itself.
- A representative perspective for the corrosive condition on the interior surface of the pipeline is obtained from the removal and evaluation of coupons.
- A later section provides precautionary notes regarding use of electronic probes (electrical resistance [ER] or linear polarization resistance [LPR]) whenever the probe elements are in environments containing H_2S. The H_2S reacts with iron to form highly conductive iron sulfide, which can "bridge" or form electrical shorts across probe elements, invalidating data.

9.2.1 Determining a General Weight Loss Corrosion Rate

NACE SP-0775-2005 or the most current year, provides equations for determining general weight loss corrosion rates.[54] The basic formula for the corrosion rate in mils per year or in millimeters per year is

$$CR = F \times [W/(A \times T \times D)] \quad \text{mpy or mm/y,}$$

where

F = Different factor, depending on units of input and whether output is in mils per year (mpy) or millimeters per year (mm/y)
W = Mass loss in grams (g)
A = Initial exposed surface area (square millimeters or square inches)
T = Exposure period in days
D = Density of the metal coupon (typically grams per cubic centimeter (g/cm^3)

The factor "F" will be different, based on whether the surface area of the coupon is measured in square inches or square millimeters.

- To report corrosion rates in mils per year, use $F = 2.227 \times 10^4$
- To report corrosion rates in millimeters per year, use $F = 3.65 \times 10^5$

[Those factors are built by the combination of conversion factors to match grams to grams, convert inches to centimeters, millimeters to centimeters, inches to mils, and days to years. One inch equals 1,000 mils. The dimensions "cancel" each other, yielding corrosion rates in mils per year or millimeters per year.]

When calculating general corrosion rates, make the fundamental assumption that the exposed surface area remains constant, and the only thing changed is the relative thickness of the coupon.

9.2.2 Example: Determining a General Weight Loss Corrosion Rate

Consider a carbon steel coupon that had a 0.0510 gm loss over a 92 day exposure. Also assume the steel coupon (and pipeline steel) has a density of 7.86 g/cm^3. For this example, assume an exposed surface area of 3.40 in^2. The goal is to determine corrosion rate in mils per year, where 1000 mil = 1 in. Thus, per the equation above, the corrosion rate is calculated as follows:

$$CR\ (\text{mil/y}) = \frac{F \times (0.051\ \text{g})}{[3.40\ \text{in}^2 \times 92\ \text{days} \times 7.86\ \text{g/cm}^3]} = 0.46\ \text{mpy.}$$

The calculated corrosion rate of 0.462 mil/y is below the generally recognized maximum acceptable corrosion rate of 1.0 mpy, as presented in NACE SP-0775-2005.[54] The corrosion rate calculations in millimeters per year would follow the same equation, except substitute the metric factor F and metric measurements for surface area in the calculations.

Sidebar: Illustrating the Basis for the Corrosion Factor "F"

The numbers that are used when determining the factor "F" are highlighted in bold font:

$$CR = \frac{(0.0510\ g) \times \mathbf{(1\ in/2.54\ cm)} \times \mathbf{(1000\ mil/in)}}{([3.40\ in2 \times \mathbf{(2.54\ cm/in)^2}]) \times [92\ days \times \mathbf{(1\ year/365\ days)}] \times [7.86\ g/cm^3]}.$$

The factor F, which is necessary for the corrosion rate to be in mils per year, is

$$F = \frac{(1\ in/2.54\ cm) \times (1000\ mil/in)}{(2.54\ cm/in)^2 \times (1\ year/365\ days)} = 22{,}274 \approx 2.227 \times 10^4.$$

9.2.3 Determining Pitting Rate, Based on Evaluation of Coupon Removed from System

When corrosion coupons are placed within pipelines, pits may form on the metal coupons, just like they form on the base metal of the pipe. The weight loss corrosion coupons can also be used to detect and measure pitting corrosion rates. However, as opposed to measuring the weight or mass loss, the depth of individual pits on the metal coupon are measured, and pitting rates are determined based on the depth of the deepest pits over the exposure period.

The analysis of internal corrosion monitoring starts with a visual inspection of every coupon removed from service for evidence of any pits. Then, if pits are observed, the follow-up activity is to determine what caused those pits to form. There are a number of potential corrosion mechanisms, which could have generated the pitting. These range from microbiological activity, to the presence of carbon dioxide (which reacts with water to form carbonic acid), oxygen ingress, or perhaps the presence of any hydrogen sulfide (H_2S). The clues may come from analysis of the corrosion products or deposits on the coupons, or perhaps results from samples of the fluids being transported through the pipeline. Then, once completed, those deposits are cleaned from the metal surface, enabling an assessment of pitting or other corrosion mechanisms.

The depth of individual pits can be measured either through the use of a light microscope or by using a needle gauge. The key is to measure the depth of an individual pit, and compare that reading to one taken on unaffected base metal. The difference is the depth of the individual pit. Although the depth of the deepest pit is of the greatest interest, it should also be noted whether the pit was an isolated pit, or if there are numerous pits in close proximity, such as a network of pits. The latter will be used in pipeline strength calculations.

Unlike laboratory work, the analysis of pitting observed on metal coupons must rely upon basics. The date when the coupon was installed, when it was removed, and the depth of the deepest of all the pits are known. The pitting rate would be calculated from the measured depth of the pit divided by the exposure period. The results would then be converted into an annual rate, such as mils per year or millimeters per year, as is used by NACE SP-0775-2005.[54]

Note that it is generally not possible to know the specific date any individual pit or pits initiate and start forming. Hence, the practice is to calculate the pitting corrosion rate, based on the date of installation. The validity of this practice is supported by scanning electron microscopic (SEM) investigations of specially prepared electron microscope (EM) coupons, which will be discussed in Section 9.2.5. (The researchers were able to use a SEM and observe the early stages of pitting on their highly polished coupons after just two hours of exposure to a bacterial test solution.)

Another specialty technique, which is used for determining when pit initiation occurs, is based on using a highly sensitive electrochemical noise (ECN, also EN) to detect a specific pattern of electrical noise that fingerprint pit initiation. (See also Section 9.3.3.) This is a data-intensive monitoring technique. However, ECN corrosion monitoring is primarily used in laboratories or for specialty diagnostics of piping or process equipment. Multiple data points are collected every second, necessitating sophisticated data collection and processing equipment. Skilled specialists are also required for consistent evaluation and interpretation of the results. As such, the use of EM coupons or LPR probes for detecting pit initiation is best when reserved for diagnostics or specialized applications.

9.2.4 NACE SP-0775 Categorizes Corrosion Rates

NACE SP-0775-2005[54] provides an industry consensus definition for acceptable general and pitting corrosion rates. Those values are presented in the referenced NACE document, which categorized the corrosion rates as "Low," "Medium," "High," and "Severe." These values are duplicated in Table 6.

Table 6 Categorizing Carbon Steel Corrosion Rates for Oil Production Systems

	Average Corrosion Rate		Maximum Pitting Rate	
	mm/y	mpy	mm/y	mpy
Low	<0.025	<1.0	<0.13	<5.0
Moderate	0.025–0.12	1.0–4.9	0.13–0.20	5.0–7.9
High	0.13–0.25	5.0–10	0.21–0.38	8.0–15.
Severe	>0.25	>10	>0.38	>15

mm/y = millimeters per year
mpy = mils per year (1,000 mils = 1 inch)

Source: NACE SP-0775-2005[54]

Use this table as a guideline for defining acceptable and unacceptable corrosion rates. In general, the worst or highest general or pitting corrosion rate for any individual corrosion coupon should be used to categorize the observed internal corrosion.

If the categorization is above acceptable limits, use that as a basis for adjusting chemical treatments or implementing other measures to mitigate corrosion.

9.2.5 EM Coupons for Electron Microscope Examinations—For Specialized Analyses

Weight loss coupons are used to measure internal corrosion rates and pitting corrosion rates. The general corrosion rates are determined, based on the change in mass, which occurred during the exposure period between installation and removal. Pitting corrosion rates are determined, based on the depth of the deepest pits that formed during the exposure period. An academic question has been around for years: how long is the period from the time of installation until the pit or pits initiate? Since data is not generally available to document the time associated with pit initiation and propagation, the practice has been to assume the pit formed uniformly over the exposure period. However, studies of EM coupons, which are discussed in the next paragraphs, evidence corrosion damage and the formation of pits within hours to a week after exposure in test environments, thus validating that assumption.

Although weight loss coupons are the workhorses for internal corrosion monitoring, a new tool has been developed, and now serves as a diagnostic tool. "EM coupons" are metal coupons similar to weight loss coupons, except they have been highly polished for scanning electron microscopic analysis to determine the causative mechanism of any corrosion that may have occurred after those coupons were placed into pipeline or process systems. The EM coupons are inserted into the systems of interest for very brief periods of time, such as hours, rather than the typical 90 days allocated for weight loss coupons. They are then removed for detailed examination, using scanning electron microscopy, which is used to characterize the corrosion mechanisms.[29, 58]

The analyses of EM coupons, such as shown in Figure 48, are manpower intensive, and require the skill of specialists. Hence, the methodology is comparatively expensive, and should only be used to conduct specialized analysis to fingerprint and diagnose unique problems. EM coupons should not be used as part of a routine corrosion program.

The use of EM coupons is noted here to alert the reader that additional resources and tools are available for specialized diagnostics, if and when needed.

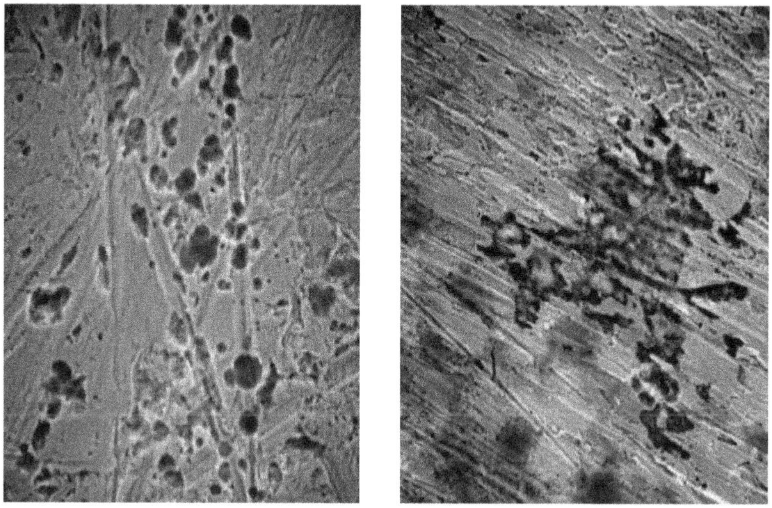

Figure 48 EM Coupon for Specialized Diagnostics Using Scanning Electron Microscope[58]
The left photomicrograph illustrates a polished coupon after two hours exposure to a bacterial test solution. The early stages of corrosion initiation are visible. The right photomicrograph illustrates a coupon exposed to the bacterial test solution for a week. A larger pit is visible.

9.3 ER and LPR Probes

There are other internal corrosion monitoring techniques, including electrical resistance (ER) corrosion probes or linear polarization resistance (LPR) corrosion probes. The text below provides an overview for using ER and LPR probes. Note, however, that the elements of ER or LPR probes are subject to "bridging," which is an "electrical short" due to the deposition of highly conductive iron sulfide corrosion products that form electrical shorts either between different areas of a single element probe or between two or three probe elements. Bridging compromises the validity of measurements and may limit the use of electronic probes in H_2S-rich environments.

9.3.1 Overview of ER Probes

ER probes are the most common type of electronic corrosion probe used within the petrochemical industry. They can provide quick measures of the corrosion rates, based on comparing measurements of the electrical resistance of the probe between the present and previous readings. There are various designs of ER probes that can provide higher sensitivity, but have a shorter service life. Some of the more sensitive probe elements, such as a strip or tube loop, may require a velocity shield, so that the production flow does not set up a vibration, causing the probe element to fatigue and fail prematurely.

Conductive deposits, such as iron sulfide or black powder, can electrically "short" across a probe element, causing invalid results until those corrosion products are cleaned and removed. The cylindrical ER probe design shown in the middle of Figure 49 is the least susceptible to iron sulfide deposition forming a short or bridge across the probe element.[59]

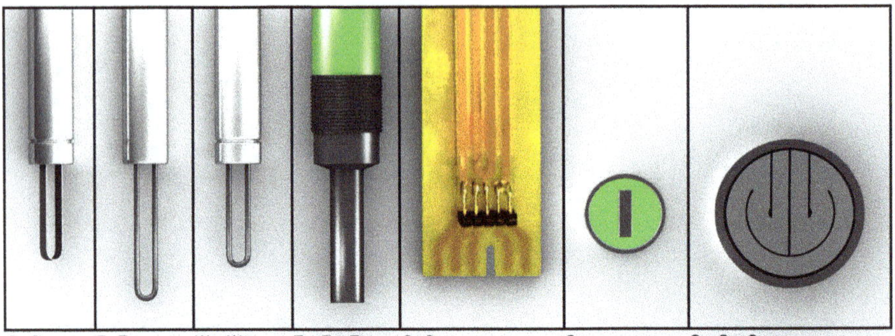

Figure 49 Different Styles of Electrical Resistance ER Probes[59]
Photograph courtesy of Rohrback Cosasco Systems

9.3.2 Overview of LPR Probes

LPR probes, such as the one depicted in Figure 50, are based on a different electronic technique, which makes each reading independent of all previous readings. Here is how LPR probes work: The potential difference between two probe elements is measured, and then a current is applied to force a small shift or increase in the potential differences, such as 10 mV. The applied current that was necessary for forcing the (10 mV) potential increase is measured, and that current is then used to calculate the instantaneous corrosion rate.

LPR probes are best for measuring corrosion rates in aqueous systems, such as water injection systems, but typically have limited use in systems transporting hydrocarbons.

There are two primary concerns with LPR probes. The first is that the LPR technique is best when used in a conductive medium, whereas oil and gas production is considered a low conductivity application. The second is the potential bridging of electrodes when conductive iron sulfide deposits form and extend between the elements of the LPR probe. Such bridges may form whenever both water and H_2S are present. The conductive iron sulfide deposits provide electrical shorts between the probe elements, functioning much like the electrolyte in a corrosion cell. The result is that once the bridge forms between the probe elements, future measurements would be invalid until the probe elements are cleaned and the bridge has been removed.

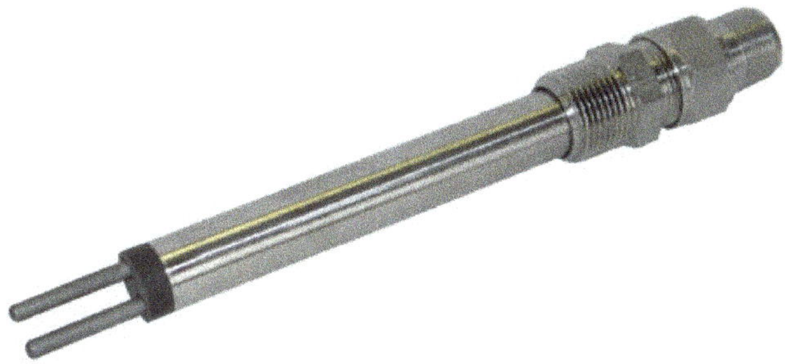

Figure 50 Two Electrode Linear Polarization Resistance Probe
Photograph courtesy of Rohrback Cosasco Systems

9.3.3 Overview of Electrochemical Noise Monitoring

A variation of the LPR monitoring is the electrochemical noise (ECN) monitoring. Electrochemical noise refers to the electrical pulses generated when pits form and propagate. Like LPR readings, every ECN measurement is independent of all previous readings. The time interval between each ECN reading must be shortened to seconds or even fractions of a second, rather than minutes or hours, which are more typically the interval between LPR readings. Otherwise, it would be highly unlikely to detect the electrochemical signals characteristic of corrosion mechanisms. Thus, sophisticated telecommunications systems are needed to collect and record the large volumes of electrochemical data necessary to detect the corrosion mechanisms. In addition, skilled corrosion professionals must be available to be able to review and interpret those results. Although the use of ECN may be practical at certain processing facilities, it is unlikely to be cost-effective in cross-country gathering or transmission pipelines. Hence, the use of ECN techniques is typically reserved for specialized analyses.

9.4 Sulfide Bridges May Form Electrical Shorts and Compromise Monitoring

CAUTION: If iron sulfide deposits onto or across the elements of electronic corrosion probes, those highly conductive depositions will cause electrical shorts, which may result in inaccurate measurements of the corrosion rates. Consider the design of the electrical resistance (ER) probes previously depicted in Figure 49. The strip loop, tube loop, and wire loop on the left side of the figure, plus the flush mounted probe on the far right side of the figure require the measurement of the electrical resistance be across or along the entire sensing element. Figure 51 depicts a tube loop and flush mount ER probe that have simulated iron sulfide deposits. The normal electrical path used for measuring the probe's resistance is shown on the left for each set of probes, while the shortened electrical paths caused by the depositions are depicted on the right. Whenever iron sulfide deposits form an electrical short between different parts of the sensing element, that ER probe cannot be used for subsequent measurements of the electrical resistance and corrosion rate calculations until the probe has been replaced.

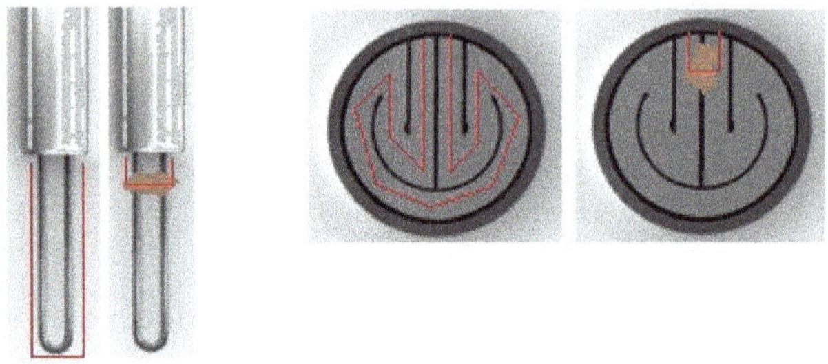

Figure 51 Illustration of Bridging Shows a Shortened Electrical Path
Photograph courtesy of Rohrback Cosasco Systems

> The red lines superimposed on the photograph are to illustrate the shortening of electrical path when electrically conductive deposits form on the probes, compromising measurements
>
> - The normal electrical path for resistance measurements of *tube loop probes* is shown on the far left, while the shortened path is on the left center.
> - The normal electrical path for *flush mounted probes* is shown in red on the right center for the second set, while the shortened electrical path is illustrated on the far right.
> - Such depositions can cause significant changes in the electrical resistance measurements, and compromise measurements.

Similarly, if iron sulfide deposits between the elements of two or three element linear polarization resistance (LPR) probes, it will not be possible to measure the voltages and currents between the elements of those probe elements and calculate the corrosion rates. Once the iron sulfide forms shorts between the probe elements, those probes need to be removed and replaced.

If iron sulfide starts to deposit between the probe elements, expect the probe readings to become erratic. Any subsequent readings should be assumed to be invalid. Although it might be possible to reactivate a probe that can be removed, cleaned, and returned to the system, do not depend on its reliability. In addition, subsequent depositions may again short the elements of any replacement ER or

LPR probe. This cycle will continue until the source of the H_2S, which forms the iron sulfide, has been mitigated and stopped.

One such approach to controlling and ideally eliminating any H_2S from entering the system would be through the continuous application of a H_2S scavenger at a dosage rate, based on a multiple of the concentration of H_2S entering the system. The multiple is necessary, based on inefficiencies in the chemical reactions that neutralize the sulfides. Check with the chemical product and service companies for their recommendations, which will take into consideration the development of any new scavenger products. Otherwise, a general recommendation is to apply 5 ppm of a H_2S scavenger for each part per million of H_2S.

9.4.1 Cylindrical Intrusive or Rectangular Flush ER Probes—Less Affected by Bridging

One corrosion probe manufacturer recommends using single element cylindrical electrical resistance (ER) probes, such as shown in the middle of Figure 49, whenever using intrusive electronic probes to monitor corrosion within systems having a history of H_2S and iron sulfide formation.[59] ("Intrusive" refers to corrosion probes that extend into the product stream within a pipeline, as opposed to flush mounted corrosion probes, which are installed to be even with the interior of the pipe wall.) The design of the flush mounted strip, which is second from the right in that figure and has a rectangular appearance, is also less susceptible to erroneous readings due to sulfide bridging.

Whenever sulfide bridging is a concern, consider the use of weight loss corrosion coupons, which can be removed and visually inspected. They would provide a good perspective regarding conditions along the interior of the pipeline. Flush mounted coupons, which are installed and adjusted to be even with the interior wall of a pipeline, also have the advantage in that they will not impede the passage of a maintenance or cleaning pig.

9.4.2 Summary of Strengths for Monitoring Corrosion When H_2S is Present

Corrosion coupons and electronic probes are tools for measuring corrosion rates within potentially corrosive systems, such as pipelines. Corrosion coupons are selected to match the composition of the base metal of the pipe wall, and as such are a representative measure for assessing conditions on the pipe wall. Many designs for the ER electronic probes are subject to an electronic shorting when iron sulfide deposits on the probe element(s)—particularly the wire loop or tube designs or the flush mounted designs—where the sensing element is folded or turned to achieve the longest continuous length within the smallest surface area (for sensitivity).

LPR probes are also subject to bridging between elements. Consequently, weight loss coupons or ER probes, which have a cylindrical or rectangular shaped element, are generally the most practical monitoring tools for measuring corrosion rates within pipeline systems where H_2S is present.

9.5 Proper Placement of Coupons or Probes for Optimum Monitoring

The following sections present the proper placement of either corrosion coupons or corrosion probes within pipelines, so that valid results are obtained. There are three considerations, namely (a) the placement of the coupon near the end of the line being monitored, which can be used to ensure adequate corrosion inhibitor treatments are being employed; (b) the placement at the proper azimuth or clock orientation, such that the probe element is wetted by any electrolytes flowing through the pipeline; and (c) considerations for solids, such as sand or black powder, which could abrade the pipe wall or collect in the small gap between a coupon or probe holder and the hole cut to provide the access for the corrosion monitoring. If, for example the hydrocarbon production is accompanied with small volumes of formation fines, such as sand, there may be a potential for those solids to gall the threads of the monitoring access fitting whenever the monitoring coupons or probes are exchanged. Hence, monitoring locations may need to be adjusted to minimize the potential for sand or other solids.

Section 9.5.1 will present the placement of coupons or probes for measuring corrosion rates within a pipeline segment, and would also be used to confirm the adequacy of corrosion control mitigation treatments. Section 9.5.1.1 will include a brief discussion regarding sampling to analyze the concentration of H_2S near the beginning and end of a pipeline segment. The difference in the concentration serves as a metric, which can be used to determine the rate of formation of black powder within individual pipeline segments. Section 9.5.2 provides illustrations regarding the proper orientation for such monitoring, so that valid results are obtained. Section 9.5.3 illustrates an approach for orienting the access fitting, so the flow will continuously flush solids away from the monitoring or sample point itself.

9.5.1 Place Coupons or Probes Near the End of a Pipeline

This section provides an overview for ideal locations to place internal corrosion monitoring and sampling. These locations could also be used for collecting samples of the process fluids and gas if sample ports have been installed, or samples can be collected through the access fittings.

9.5.1.1 Monitoring for a Single Pipeline

Figure 52 illustrates a single pipeline from producing wells, to a gathering or processing facility. Two monitoring points are shown: one at the start of the pipeline, and one near the end of the pipeline. The upstream location near the start of the pipeline can be used to monitor the corrosion rates or corrosiveness of the fluids and gas entering the pipeline, prior to any treatment. The downstream monitoring point–the one nearest the end of the pipeline segment–is used to measure the downstream corrosion rates and ensure the adequacy or effectiveness of any chemical treatments or other mitigation measures.

Results from the downstream monitoring would be used as the basis for adjusting any chemical treatments, such as corrosion inhibitors. The theory is that if corrosion inhibitor products are injected at the beginning or start of the pipeline, those products should be able to protect the entire length of the pipeline. However, excessive concentrations of corrosion inhibitors or other chemical treatments could potentially upset downstream processing equipment, such as separators. Hence, any chemical treatments should be just enough to protect the pipeline.

The chemical injection rate at the beginning of the pipeline is slowly increased through a series of step changes or increases, until adequate internal corrosion protection is obtained at the end of the pipeline. This is based on having acceptable corrosion rates (see Section 9.2.4), as measured at the end of the pipeline segment, and keeping the chemical treatments as low as possible, such that the concentration of the chemicals applied does not affect the functionality or operation of any downstream process equipment, e.g., separators, which are used to separate gas, hydrocarbons, and any produced water.

A generic term used to identify transient operating conditions outside of the normal operating conditions is "process upset." Corrosion control professionals must work hand in hand with operations personnel to minimize process upsets, while at the same time striving to maintain the integrity of the pipelines and process equipment. (One example of such a partnership was in the selection of a corrosion inhibitor used for protecting a crude oil transmission pipeline. This was to ensure the inhibitors within that crude oil pipeline did not adversely affect the catalysts, which were used in downstream refinery operations.)

If process upsets are encountered, consult with a technical representative from the chemical product and service provider to consider other corrosion inhibitor products. The testing methodology, which is used to establish the required chemical treatment rate, is presented in greater detail elsewhere.[60–61]

The key is to select the corrosion inhibitor products that offer the best cost performance, wherein the cost performance is based on the price of the product, including concentration, the required treatment rate, and the expenses to deliver the product to the chemical injection site.

A general rule of thumb is to seek a 10% increase in cost performance before considering the substitution or replacement of any chemical product. Factors to consider include compatibility with other chemical products used to treat the pipeline or pipeline network; logistical support for delivering the products; taking advantage of any volume discounts for chemical purchases; and ensuring compatibility of the as-delivered (new) products, with the previous product to prevent the formation of emulsions, which could clog up pumps and the associated chemical injection lines if the previous product is not compatible with the new product.

Still another consideration for single pipelines is that the concentration of H_2S has to be measured at the beginning of the pipeline, as well as at the end of the pipeline. The sampling and field analyses for the comparative concentrations of H_2S could be through sample ports at the two ends of the pipeline, or perhaps through ports on the access fittings or the extraction tools used to insert and remove the corrosion probes or metal coupons. Compare the parts per million of the H_2S at the beginning and end of the pipeline, along with the flow rate (MMscfd or m^3/d) through the pipeline to calculate the total volume of iron sulfide that formed due to reaction of that H_2S with the iron in the pipe wall (see also Section 2.3.1).

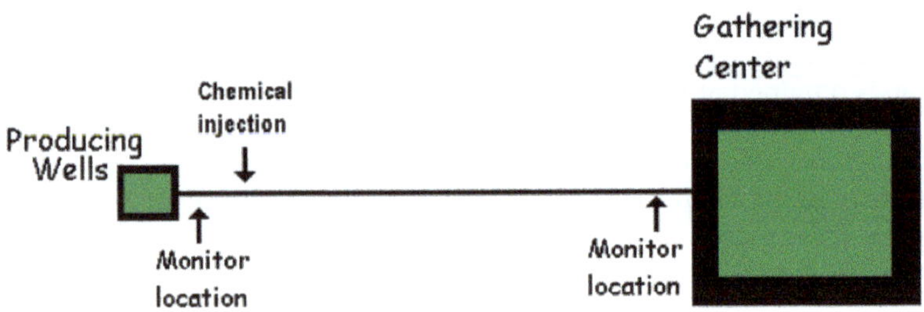

Figure 52 Use Downstream Monitoring Point for Controlling Treatments[60-61]

9.5.1.2 Monitoring for a Gathering System

The gathering system depicted in Figure 53 is slightly more complex. Break the system down to individual pipeline segments, or "legs" as they are sometimes called. Then focus on the monitoring points near or at the beginning and end of each pipeline segment. If two pipeline segments are flowing to a common pipeline, set the priorities for monitoring points at the ends for each of the two pipeline segments, which feed or flow into the common line, and then check the monitoring point near or at the end of that common line. The chemical treatments applied to each of the segments feeding the common line may need to be adjusted, such that adequate protection is also attained at the end of the common line.

Likewise, and following the example from the single pipeline above, periodically collect samples and measure the concentration of H_2S at the end of each of the two legs, and then measure the H_2S at the end of the common line. Take into consideration the flow rates (MMscfd or m^3/d) through each of the two legs to determine the total quantity of H_2S entering the common line, and compare it to the combined flow and concentration of H_2S at the end of the line. That will provide an estimate for the quantity of iron sulfide that may have formed within the pipeline.

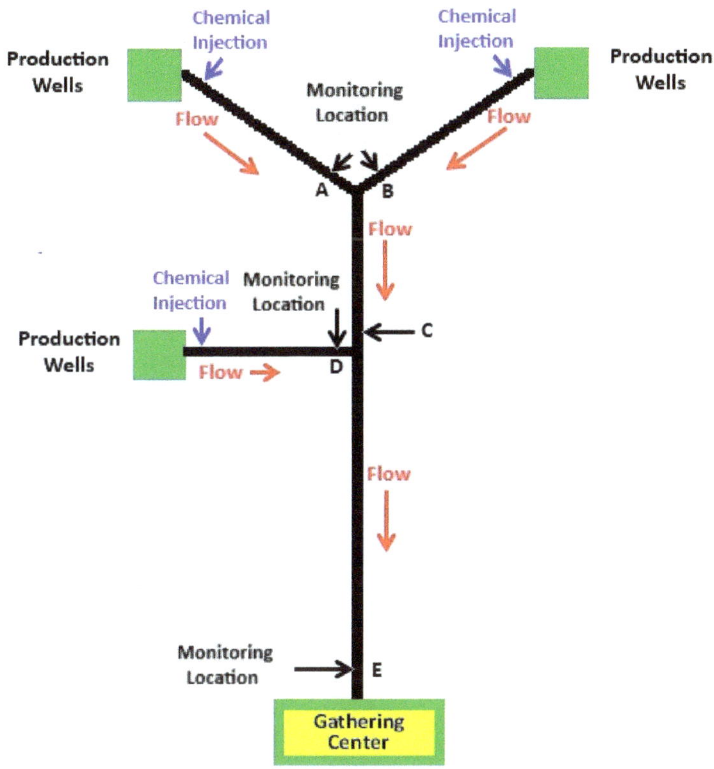

Figure 53 Preferred Monitoring Locations for a Gathering System[60-61]
The downstream monitoring points are used for controlling treatments.

Comparative measurements for the concentration of H_2S at the different sampling points throughout the pipeline gathering system (A, B, C, D, and E in Figure 53) can be used as the basis for estimating the quantity of iron sulfide forming within the pipeline. Likewise, analyses of samples collected at these points can be used to assess the effectiveness of mitigation measures, such as chemical treatments.

9.5.2 Orient Coupon or Probe Element into Electrolyte

Two papers published many years ago were written to provide pipeline designers with an understanding of flow within pipelines, because too many monitoring points were installed at the wrong orientations for providing valid, representative data.[40, 60] Unlike an assumption of uniform conditions through the pipeline, the proper orientation of the monitoring instrumentation, whether it be a metal coupon or an electronic probe, may be critical for obtaining valid results whenever there is either two or three phase flow. Two-phase flow refers to liquid and gas both flowing through a pipeline, while three-phase flow refers to crude oil or hydrocarbon, liquid water, and gas flowing through a single pipeline. The key to effective monitoring is to place the metal coupon or electronic probe in the aqueous phase, because water is the electrolyte necessary to support internal corrosion.

Consider Figure 54, which illustrates the placement of a corrosion probe within a system being monitored. Look at Figure 54(a) on the left. A corrosion probe located at the 3:00 or 9:00 o'clock orientation is not wetted by the electrolyte in multiphase flow. As such, it could not properly monitor the potential corrosion occurring on the bottom of the pipe, which is wetted by the electrolyte. (The electrolyte is part of the corrosion cell, as presented in Section 9.1.1, and corrosion can be found only where the electrolyte is or has been present.)

Next, look at Figure 54(b) on the right, which depicts an electronic probe oriented at the 6:00 position. This probe is positioned to be along the path where any electrolyte, such as produced or condensed water, would be swept, and as such would yield valid results.

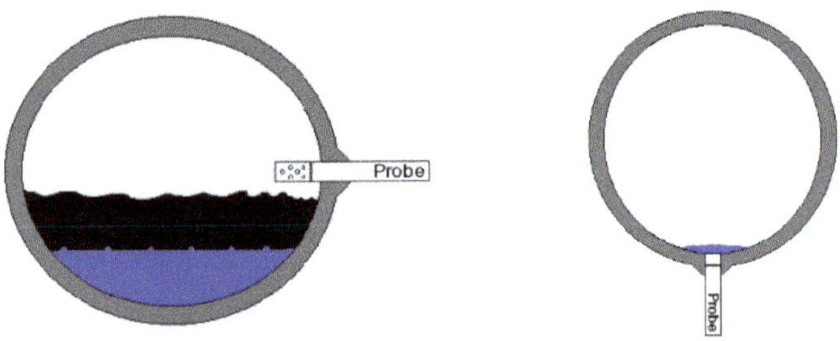

a b

Figure 54 Orient Probe or Coupon into the Electrolyte Flowing Through Pipeline[40]
(a) Probe is not immersed within the most corrosive phase, and will *not* yield valid results. (b) Probe is properly positioned to monitor corrosion rates for liquids swept through pipe. However, such locations are not recommended if sand or solids are also present and being swept through the pipelines, because those solids pose a threat for galling the threads of the access fitting.

9.5.3 Considerations When Sand or Solids Pose a Threat for Galling Fitting

Figure 55 illustrates a corrosion monitoring point on the bypass pipeline around a pig receiver within a gathering system. If the corrosion monitoring point was on the bottom of a horizontal pipeline segment, such a location could be susceptible to the buildup of particulates, such as sand, which can gall the threads within the monitoring point access fittings. However, this particular location is on the extrados (outside radius) of a 90° bend, so that the natural flow of any liquids will flush any buildup of solid particulates away from the fitting. That minimizes the likelihood for any particulates, which are within the pipeline, from falling or being swept into the interior of the access fittings. The concern is that those particles could potentially gall the threads of the fitting whenever the electronic corrosion probes or metal coupons are removed and replaced.

Figure 55 Flow Flushes Particles from Access Fitting on Extrados of 90° Bend Such a location helps sweep sand particulates and prevents the threads on access fitting from galling.

Whenever there is a possibility of localized accumulation of solids within a pipeline, both operations personnel, as well as corrosion control professionals need to be aware of potential galling of the threads on access fitting when sand gets onto those threads. Sand or quartz is a very hard material, which can cut the threads of the carbon steel access fittings, making it difficult–if not nearly impossible–to properly close and seal the ports of the coupon or electronic probe access fittings. There is a potential threat to maintaining system integrity if the access fitting cannot properly seal. Instead, the coupon or probe extraction tool might have to be kept in place to maintain pressure containment and system integrity. When such threats exist for sand galling access fittings, consider moving the corrosion monitoring locations slightly upstream or downstream to a location where sand or other particulates would be naturally swept away from the access fitting.

10

Comparing Sensitivities of Internal Corrosion Monitoring and Inspection Techniques

This chapter compares the relative sensitivity of the internal corrosion monitoring techniques using weight loss coupons and electron corrosion rate probes to inspection techniques, such as ultrasonic (UT), radiographic (RT), and magnetic flux leakage (MFL), which are best for assessing system integrity and pressure containment. The discussion starts with a quick overview of the inspection methodologies, primarily to focus on their ability to measure the depth of pipeline anomalies, here described as their sensitivity or ability to detect active corrosion. Table 7 at the end of this chapter compares the relative sensitivities to their abilities to detect the onset of corrosive conditions.

This chapter will compare the relative sensitivity of the standard inspection techniques (UT, RT, radiographic, ultrasonic, and MFL) to the standard corrosion monitoring techniques (weight loss coupons and electronic corrosion probes). This is to show the essential role of internal corrosion monitoring for early detecting and quantifying the onset of corrosive conditions, compared to the inspection techniques, such that remedial measures can be implemented to minimize potential wall loss.

These discussions will show that the use of weight loss metal coupons is perhaps the most sensitive, cost-effective methodology for quick-or early detection of the onset of corrosion. Additionally, metal coupons are less affected by depositions of iron sulfide, which are highly electrically conductive, and can compromise the functionality of corrosion monitoring instrumentation whenever those deposits form electrical shorts or bridges across or between probe elements. The weight loss coupons are perhaps the most cost-effective monitoring technique for systems where H_2S and black powder formation is possible.

Note that ultrasonic, radiographic, and MFL inspection techniques are best for overall integrity assessment, but are not the best choice for early detection of any potential internal corrosion, regardless of whether black powder is or is not a concern. Internal corrosion monitoring and the non-destructive inspection techniques provide complementary perspectives of the onset and subsequent consequences from corrosion mechanisms.

10.1 Overview and Sensitivity of Ultrasonic (UT) Inspection

Whenever a new corrosion monitoring point is proposed for an existing pipeline, one of the first questions asked is whether the same data could be obtained through ultrasonic inspections at the same location. There is natural reluctance to install a new access fitting through a "hot tap," which is the pipeline industry's jargon to describe the process for welding an access fitting onto an operating pipeline (under pressure), and then cutting a hole through the pipe wall inside the fitting) to allow access to the interior of the pipeline.

(The installation of access fittings through hot taps is a standard industry practice. Nevertheless, there is some risk associated with such operations. Although installations could be conducted during planned outages or shutdowns, competing priorities may preempt such work. The true preference is to have all monitoring points and sampling points installed during the initial pipeline fabrication.)

Perhaps the best explanation for the benefit of monitoring corrosion by the use of corrosion coupons or electronic probes, as opposed to relying upon localized ultrasonic wall thickness measurements, is through comparing the relative sensitivity of the monitoring versus the inspection technique. Ultrasonic wall thickness measurements, previously illustrated in Figure 40, are a standard tool for assessing the integrity of pipelines and/or assessing the integrity of vessels and tanks. The key word is "integrity." It answers the question regarding whether the walls are sufficiently thick for pressure containment. However, ultrasonic thickness measurements are not sufficiently sensitive for measuring and controlling potential internal corrosion.

The accuracy of such UT wall thickness measurements is typically reported as ±10 mils (0.010 in or 254 μm). However, many UT inspectors claim they can detect changes as small as ± 5 mils (0.005 in or 127 μm). Consider the time required to be able to detect and confirm a 10 mpy (254 μm/y) corrosion rate, where two readings or measurements are commonly used to confirm a corrosion rate trend. It could take between 1–2 years to be able to confirm a 10 mpy (254 μm/y) corrosion rate, using UT techniques, depending on the UT inspector's ability to detect changes as small as 5 or 10 mils (127 or 254 μm).

In contrast, weight loss coupons, as presented in Section 9.2, provide corrosion rates every 90 days, or whenever removed from the system and evaluated. Corrosion rates are typically reported to two decimal points, or a hundredth of a mil per year (0.025 μm/y). [Each style of weight loss coupons are precisely manufactured to have a very specific surface area, which can be accurately measured. The density and precise alloy of steel used to manufacture the coupons is selected to match the metallurgy of the system being monitored. Additionally, the initial and final weight or mass of the coupons are measured to four decimal places. The number of days the coupon was installed and used to monitor the system is part of the documentation. Thus, the two-decimal digit precision of the corrosion rate measurements can be justified.]

This means weight (mass loss) metal coupons offer far more sensitivity and precision for detecting and measuring corrosion rates, versus successive UT wall thickness measurements, which have a higher degree of uncertainty (± 10 mils or ± 254 μm). As such, metal coupon monitoring results should be used for evaluating the effectiveness of corrosion mitigation measures.

10.1.1 UT-ILI Pigging: Typical Sensitivity

Before leaving the topic of ultrasonic inspections, consider in-line inspection (ILI) vehicles using UT technology. UT-ILI vehicles function similarly to that of the handheld UT meters, except the UT-ILI vehicles have multiple channels, each looking at a specific orientation around the internal circumference of a pipeline. The companies providing the ILI services typically claim the sensitivity or ability to detect pipeline anomalies greater than 5% of the nominal wall thickness. Thus, for a 0.375 in (0.953 cm) nominal wall thickness, the 5% threshold would equate to 19 mils (480 μm), which is less sensitive than the 10 mil (254 μm) sensitivity for a handheld UT meter.

This underscores the true value of the ultrasonic ILI vehicles. They have tremendous data collection and processing capabilities, and are an ideal tool for inspecting the interior of a pipeline to identify the precise location of any corrosion indications, while providing a good estimate for the depth and size of those indications. It is a great tool for integrity management. However, neither UT-ILI, nor handheld UT meters, have the sensitivity for measuring corrosion rates with the precision necessary for assessing the effectiveness of corrosion mitigation measures over the short time durations necessary for controlling corrosion rates.

10.2 Overview and Sensitivity of RT Inspection

Another standard technique used for inspecting steel in pipelines or elsewhere is radiography (RT). The depth of individual indications can be seen from observations of individual radiographs, while comparisons of two successive radiographs taken from the same location but at different times would be used to estimate a corrosion rate (see Figure 56).

The approach here is quite similar to that described above for two successive UT measurements. Unfortunately, like ultrasonic wall thickness measurements, the analysis of successive radiographic readings cannot provide as accurate a measure of the corrosion rates as would be available through the use of weight coupons, which are typically based on a 90 day exposure cycle.

Like UT, most radiographers report the accuracy of RT wall thickness measurements is typically reported as ± 10 mil (0.010 in or 254 μm). However, many RT inspectors claim they can detect changes as small as ± 5 mil (0.005 in or 127 μm). Thus, like UT measurements, it would take RT inspectors between 1–2 years to be able to confirm a 10 mpy (254 μm/y) corrosion rate, using RT techniques, depending on the RT inspector's ability to detect changes as small as 5 or 10 mils (127 or 254 μm).

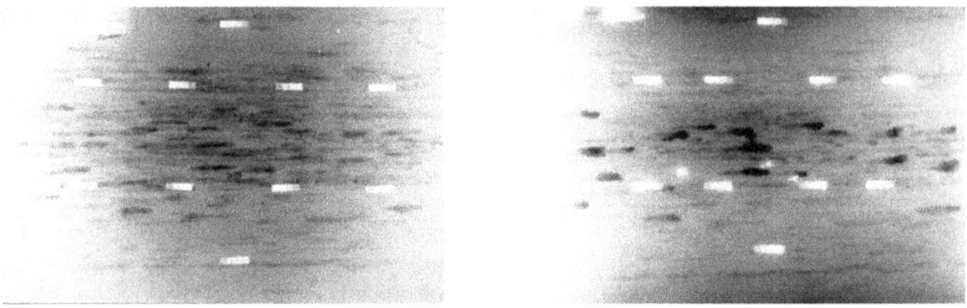

Figure 56 Comparison of X-Rays at Same Location, but Different Times
The white rectangles are metal step wedges, each having known thicknesses, which are placed on the exterior of pipe to assist radiographers determining the depth of the anomalies.

10.3 Overview and Sensitivity of MFL Technology

Still another inspection technique commonly used for inspecting pipelines or pipe segments within oil and gas production and processing facilities is magnetic flux leakage (MFL). The MFL technique is the most commonly used technique for ILIs of pipelines. It is also used for inspection of tubes within heat exchangers. Figure 39 in Section 7.1.1 described the technique.

MFL inspections, whether employed through an ILI vehicle, or through a MFL probe that is pushed or pulled through a heat exchanger tube, is an effective tool for detecting the location and approximate size or depth of anomalies, which are suspected corrosion indication prior to follow-up inspections for confirmation. The key is having a magnetic field, which saturates the steel pipe or tubing being inspected. Detectors, which are built into the inspection tool, sense changes in the magnetic field at the location of the anomalies, and are used to characterize base metal anomalies.

The magnetic fields of an MFL tool can be oriented axially, along the axial direction or length of the pipeline or pipe within a heat exchanger, or the magnetic field could be oriented in a transverse (perpendicular) direction, relative to the axis of the pipe or tubing, such that the magnetic field is aligned around the circumference of the pipe or tubing. The specific choice of orientation is based on setting the alignment of the magnetic field either perpendicular, or at an acceptable angle to the direction of suspected or known pipeline anomalies, making those anomalies easier to detect.

(The orientation of the sensors is critical for successful inspections. As an example, consider a row of marching soldiers on a field. They would be much easier to spot if the entire row of soldiers was visible, left to right. However, if observed from the side, only one could be observed. Analogously, inspection tools, which are selected for detecting known or suspected corrosion mechanisms, need to have their sensors properly oriented to be perpendicular to those known or suspected anomalies, which is to maximize the ability to detect the corrosion mechanisms of concern, such as cracking.)

The accuracy of the MFL technique is reported as a percentage of wall thickness, with 5% wall thickness being the most commonly claimed accuracy. To put that into perspective, consider a 0.375 in (9.525 mm) pipe wall thickness. The claimed precision would be 0.375 × 0.05 = 0.019 in = 19 mil (480 μm). If the actual corrosion rate was 10 mpy (254 μm/y), it could take 2–3 years to detect and confirm a 10 mpy (254 μm/y) corrosion rate using MFL techniques.

10.4 Comparative Ability to Detect Onset of Corrosion

Corrosion coupons are manufactured to have very specific dimensions, which define the surface area. The weight or mass of each coupon is measured to four decimal places. As previously discussed in Section 9.2, when metal coupons are removed from service and cleaned per NACE standards, the coupons are visually inspected and reweighed, and the corrosion rates are calculated.[54] The corrosion rates determined from the weight loss coupons are typically reported to the nearest tenth or hundredth of a mil per year (or fraction of a micrometer per year).

Metal coupons can detect a 10 mpy (254 μm/y) corrosion rate within a 30–60 day exposure period, which is much shorter than the 1–2 years required by UT or RT techniques, or the 2–3 years required by MFL techniques. Per the ER probe manufacturer's product specifications, the most sensitive wire loop electrical resistance probes can detect a 10 mpy (254 μm/y) corrosion rate in as little as four days. However, such sensitivities come at the expense of having a significantly shortened service life for those particular probes. Consequently, the most sensitive probes are usually reserved for special performance tests, while less sensitive but longer lifetime probes are used for routine corrosion monitoring. (One example of the special performance testing, which would require the most sensitive ER probes, is the performance tests used to evaluate new corrosion inhibitor products.)

The time required by UT, RT, or MFL techniques to detect and confirm a 10 mpy (254 μm/y) corrosion rate can be compared to the times required for internal corrosion monitoring by either corrosion coupons or the most sensitive wire loop ER electronic probes. This is presented in Table 7 below.

Table 7 Relative Time for Monitoring or Inspection to Detect 10 mpy Corrosion[60]

Monitoring or Inspection Technique	Time to Detect and Confirm a 10 mpy (254 μm/y) Corrosion Rate
Magnetic Flux Leakage	2-3 years
Ultrasonic or Radiography	1-2 years
Metal Coupon or ER Probe	30-60 days
Most Sensitive Wire Loop ER Probe	4 days

The bottom line is that internal corrosion monitoring is far more sensitive than external inspection techniques, such as RT, UT, or MFL. The reader is advised to use internal corrosion monitoring techniques whenever monitoring internal corrosion rates, and to use inspection techniques for integrity (strength) assessments.

The goal of the internal corrosion monitoring is to be able to detect the onset of any active internal corrosion as quickly as possible, such that appropriate remedial measures can be quickly implemented to minimize the loss of pipe wall thickness. If unchecked, such corrosion and associated wall loss could affect system integrity and pressure containment. It is important to remember once metal is lost from a pipe wall, it is gone. It is not practical to put metal back onto a pipe—welders can only address localized areas.

Know Relative Sensitivities for Measuring Corrosion Rates

- Metal coupons or ER Probes are the most sensitive technique for monitoring and controlling internal corrosion mechanisms.
- UT, RT, and MFL are inspection techniques that are best for assessing pipeline integrity.
- Early detection of internal corrosion enables the prompt initiation of mitigation measures, which can extend a pipe's service life.

It is not practical to put metal back onto a pipeline—except localized welding.

10.5 Be Alert for New Developments in Monitoring Corrosion Technology

The manufacturers of corrosion monitoring equipment have a history of developing new tools and monitoring instrumentation. One instrument that was recently introduced is a new ultrasonic-based monitoring tool, which according to the manufacturer, provides resolution down to 0.1 mil (2.54 μm). As opposed to using a single ultrasonic transducer, this tool uses many different transducers in a matrix array. The enhanced sensitivity is possibly being achieved through a combination of state-of-the-art transducers, higher frequency analog-to-digital converters, advanced data processing, and enhanced statistical algorithms. The matrix array of ultrasonic transducers may also offer advantages for monitoring internal corrosion rates from the exterior of a pipeline whenever intrusive monitoring is not feasible. However, such a tool is likely to be more expensive, considering the costs associated with new product development. Anticipate new technical papers will be published, as this technology matures.

11 Collecting Samples of Black Powder Particulates—Inside the Pipeline

Chapters 11 and 12 discuss different aspects related to particle size and the transportation of black powder particles through pipelines. For example, Section 11.1 provides a discussion on collecting samples of the black powder being transported through operating pipelines, and Section 12.3 uses knowledge of particle size for selecting the appropriate equipment to separate or remove those particles.

Many published studies of black powder depositions focus on the laboratory analyses of the samples, which are typically obtained from filter cartridges or from the maintenance pigging receiver traps. Those published studies focused on the laboratory techniques, which were used to analyze the samples. This may include discussions of the solvents used to strip away any hydrocarbons or oils encapsulating the particulates, allowing chemical analysis of those particulates. The field sampling technique presented in Section 11.1 is different from traditional sampling gathered from filter cartridges of maintenance pigging in that it focuses on collecting particulate samples from the interior of a pipeline. That enables a more representative analysis for the distribution of size of the particles. Knowledge of the true particle sizes makes it possible to determine whether particles, such as black powder, will be transported or may accumulate.

It is possible to collect samples onto special nylon filters in a matter of minutes, and to then determine the size of those particles, including whether the particles have agglomerated into larger sized particles or not. This is a very brief operation, only taking a few minutes. Longer sampling cycles would result in multiple layers of particles being collected on a nylon filter, and that would make the determination of particle size more cumbersome. Additionally, the agglomeration of particles would make it much more difficult to measure individual particle sizes. Nevertheless, a longer sampling period, e.g., 15–60 min is essential for making accurate estimates of mass flow rates for all particulates being transported through the pipeline.

Knowledge of the true size of the particulates passing through the pipelines may also help operations personnel to (a) plan the cleanings of any upstream separator vessels, such as slug catchers, or (b) separators, which have been specifically designed for removing particles of the sizes encountered within their process facilities. The results from particle size studies should also be compared to results from any previously collected, historical samples. Any changes should be observed and reported. For example, determine whether particles may be sticking together, or agglomerating, to form larger sized particles. The larger sized particles are typically easier to separate. Note also that the adhesive forces, which help to bind particles together, are often associated with liquids, such as glycols used in gas dehydration.

Samples collected from gas gathering systems in Pennsylvania may include particles of salt, while samples collected from offshore gas gathering systems in the Gulf of Mexico might contain formation fines, including sand. Treat each sample as a unique characterization of the pipeline system from which that sample was collected.

Collect Samples to Determine the Actual Size of Black Powder Particles

- Better selection and operation of the separation equipment
- Determine whether particles are being transported or are accumulating

11.1 Field Technique to Collect Samples for Measuring Particle Size

Intrusive sampling devices are inserted through existing ports or pipeline appurtenances and used to collect samples of the particulates flowing through a pipeline. They are inserted into the pipeline through an access fitting, similar to the fittings used for inserting coupons or electronic corrosion probes into the process stream. Once inserted, the valves are opened, and a portion of the pipeline's flow will be directed through the sampling device while the sample is being collected. Figure 57 depicts the sampling device recently used for particulate size determination at a gas processing facility along the Gulf of Mexico.

This sampling device is inserted through a pipeline appurtenance having two flanges above a valve, such as depicted in Figure 57. The sampling device is attached through the top flange, but the sampling tube is not lowered into the pipeline until the valve within the pipeline appurtenance is opened. Although partially obscured by the pipeline operator's legs, that valve is located just to the right of the operator's ankles. Once opened, the inner sampling tube, which is sometimes called a "stinger," can be positioned anywhere from the top to middle to bottom of the pipeline segment. The specific location is achieved by adjusting the hand wheel at the top of the device.

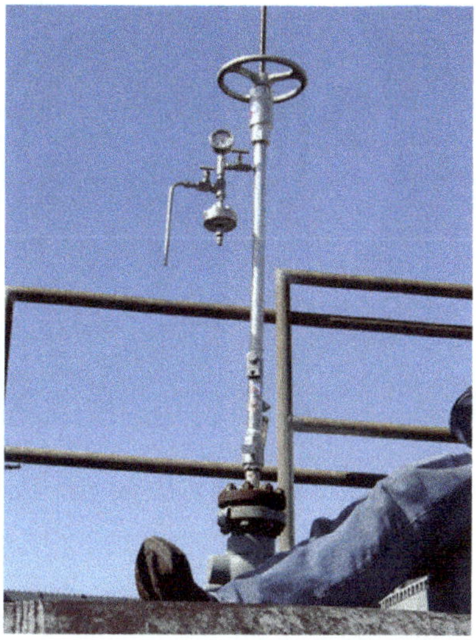

Figure 57 Test Apparatus Used to Collect Samples for Measuring Particle Size

The needle valve, which comes off and to the left of the central shaft of the sampling probe, controls the flow to the sampling filter or vent line. The pressure gage measures the pressure on the opposite side of the needle valve. The device housing the nylon filters is immediately below the pressure gage. There is also a bypass line to allow depressurizing the sampling device, which can be seen on the far left of the photo.

The pipeline pressure is known, or can easily be measured. The needle device is adjusted to achieve a specific pressure differential, which provides the driving force for the collection of particulates on the nylon filters within a reasonable sampling period. Through an iterative process, a 200 psi (1379 kPa) pressure differential was chosen, such that particulate samples could be collected in a short interval, such as 15 min.

The flow velocity can then be determined by knowing the pressure differential and the inside diameter of the sampler tubing. The filters were selected to collect samples in the range from 0.1–1.0 μm, which is the typical range reported for particles of iron sulfide. Nylon filters were selected for their mechanical durability. Figure 58 illustrates the solids collected during one particle sizing test.

Figure 58 Particles Captured on Nylon Filter for Assessing Particle Sizes

The above describes one device for collecting samples for evaluating particle sizes. However, this is not the only tool for such sampling. Others describe a slightly different device, which has also been designed to collect samples from within pipelines.[62] The most notable difference is that the second sampling device has a small flow meter built into it. The particulate samples are collected when the velocities flowing through the sampling device match the velocities within the pipeline, and this is termed "isokinetic sampling."

11.1.1 Use Intrinsically Safe Electronic Meters Where Appropriate

Sampling devices typically used to collect samples of particulates flowing through pipelines should *not* have electronic gas flow rate meters. Most facilities will not allow such electronic devices within the field or at production facilities, unless they are intrinsically safe. The concern is the possibility of electrical sparks associated with the operation of such devices in environments, where small leaks or releases of natural gas may occur. Ensure monitoring instruments or devices used at the sampling points are intrinsically safe or have been approved for operation.

11.1.2 Check for Presence of Hydrocarbons That Bind Fine Particles Together

Others, who reported their analysis of black powder collected from sales gas transmission pipelines in Japan, noted the critical need for particle size distribution analysis.[63] They collected samples into polyethylene bottles from filter cartridges and pig receiving traps throughout their network. Very sophisticated laboratory instruments and methanol extractions were then used to characterize the composition of the compounds adhering to the internal surface of the pipeline. The black powder they reported was predominantly magnetite (Fe_3O_4), with only a trace of iron sulfides. The low quantity of sulfides they reported makes sense, because their nominal gas composition only had < 20 ppb of H_2S. Other products identified from the methanol extracts of the black powder include diethylene glycol (DEG), triethylene glycol (TEG), plus elemental sulfur, which was found in a few samples. The DEG and TEG are associated with gas dehydration. They also postulated earlier formation of iron sulfides, which were subsequently oxidized to iron oxides.

Still another study reported the contaminants in the black powder sludge removed from a sales gas pipeline included "high amounts" of TEG, as well as detectable levels of elemental sulfur, sand, salts, water, and metal debris.[16] They investigated the corrosion rate of carbon steel immersed in black powder sludge, which contained high amounts of TEG, after a 355 day exposure in an oxygen-rich environment. They also reported results from a 100-day exposure in a simulated sales gas environment. The carbon steel samples immersed within the sales gas sludge showed low or extremely low corrosion rates from 0.0–0.7 mpy (0.0–17.8 μm/y) compared to blank carbon steel samples immersed in condensed deionized water, which were in the range of 1.0–4.0 mpy (25.4–101.6 μm/y). Thus, they concluded the black powder sludge had an inhibiting effect on the internal corrosion of those steel samples. This paper also reported an additional benefit in lowering corrosion rates when < 2 ppm H_2S was present.[16] The lower corrosion rates were attributed to the formation of a thin protective layer of iron sulfide. Numerous small, shallow pits on one metal coupon was covered with the black powder sludge, but which was kept within simulated atmospheric conditions (3.1 psia (31.4 kPa) O_2, and 0.006 psia (0.041 kPa) CO_2). Their observation suggests that when not in operation, gas transmission pipelines, which may contain black powder sludge, should be blanked with natural gas rather than open to the atmosphere.

Analysis of black powder sludge often results in the identification of DEG and TEG. These hydrocarbons are probably associated with upsets associated with gas processing and dehydration operations. The DEG and TEG can wet black powder particles together, such that small particles could adhere together to form larger particles. One possible advantage for making these particles larger is that they may be easier to capture.

11.2 Data to Collect When Gathering Samples for Measuring Particle Size

When collecting samples for measuring the distribution of particle sizes, it is recommended that a few additional field sample measurements should be made to obtain a more comprehensive knowledge of the localized environment within the system. The additional measurements may help later when assessing the potential threats to the system integrity.

11.2.1 Measure CO_2, H_2S, and Moisture When Collecting Samples of Black Powder

Operations or corrosion control professionals should sample the natural gas within the flow stream to measure (1) the carbon dioxide content, (2) the hydrogen sulfide content, and (3) the moisture content of the gas. Carbon dioxide could react with the steel pipe walls to form iron carbonate if any produced or condensed water is present. If hydrogen sulfide is present, there is a possibility for forming iron sulfide, the most distinctive component of black powder. The measurement of the moisture content of natural gas flowing through the pipeline will provide insight regarding the presence of water, which is the electrolyte for the corrosion cell. These are quick field tests, which should be conducted on site, and should not take more than a few minutes to complete (see Sections 5.1.1–5.1.3).

11.2.2 Consider Extraction from the Sample to Check for Dormant SRBs or APBs

If the pipeline has a history or potential concerns related to possible microbiologically influenced corrosion (MIC), consider collecting two additional small samples of the black powder sludge. Place one of those samples into a culture media solution for sulfate reducing bacteria (SRBs) and one for acid producing bacteria (APBs). Then shake each of those media solutions. This process will extract any dormant SRBs or APBs into the media solutions, which can then be used for conducting serial dilution studies for assessing microbiological populations.[47] Such data would be used to determine whether (or not) it might be appropriate to consider a biocide treatment.

The preparation and serial dilution of the sample should only take a few minutes, although the sample bottles will need to be incubated for the standard 14 or 28 days (SRB or APB, respectively) before determining a positive or negative result. Note that these proposed serial dilution studies are not the only techniques that could be used to assess potential dormant SRBs or APBs. Consult a microbiologist for other options.

These particular tests use small and unquantified volumes from the black powder sample to determine whether the sample contains dormant SRBs or APBs. Accordingly, such analysis should only be used as a positive or negative indicator for the presence of those microorganisms.

11.3 Particle Sizes and Removal of Those Particles

In Section 11.1, methods for collecting samples of black powder from pipelines were discussed, since particle size and the gas velocity directly relates to the transportation of black powder through pipeline networks.

The focus now switches to particle size, and using those sizes for identifying process equipment, which could be used to remove those particulates. This is illustrated in J.H. Perry's *Chemical Engineers' Handbook*, figure titled "Particle Dynamics,"[64] an industry standard reference, which others also cite.[65] Perry's figure is significant as it best correlates both particle size and the generally recommended process separation technology.

Perry's Figure 113 (copied here as Figure 59), provides a spectrum of particle sizes from 0.001 μm up to 100,000 μm.[64] Categories for the particle sizes include dust, mist, smoke, etc., which correspond to the full range of particle sizes reported for black powder.

Particle sizes are commonly divided into categories, such as presented in Table 8.

Table 8 Categories of Particle Sizes[66]

Category	Size
Course particles	10 μm or greater
Fine particles	Less than 10 μm
Smoke (e.g., cigarette)	Particles from 0.25-1.0 μm

Although black powder particles may form as large or coarse particles, black powder is brittle, and will quickly break into submicron size when transiting through pipelines. Thus, measures to remove black powder from processing facilities should be based on the type of processing equipment needed to remove that size particulates from the flow.

Figure 60 is an extraction from Figure 59, which illustrates commercial equipment typically recommended for removing particles from gas.[64] For example, 0.5 μm particles of dry black powder would best be collected and removed from systems using cloth collectors, such as illustrated in Figure 3. However, if the black powder particles were wetted by hydrocarbons, such as triethylene glycol (TEG), which is a carryover from gas dehydrators, or lube oil from compressors, those fine particles of black powder would typically agglomerate or stick together. That would

PARTICLE DYNAMICS

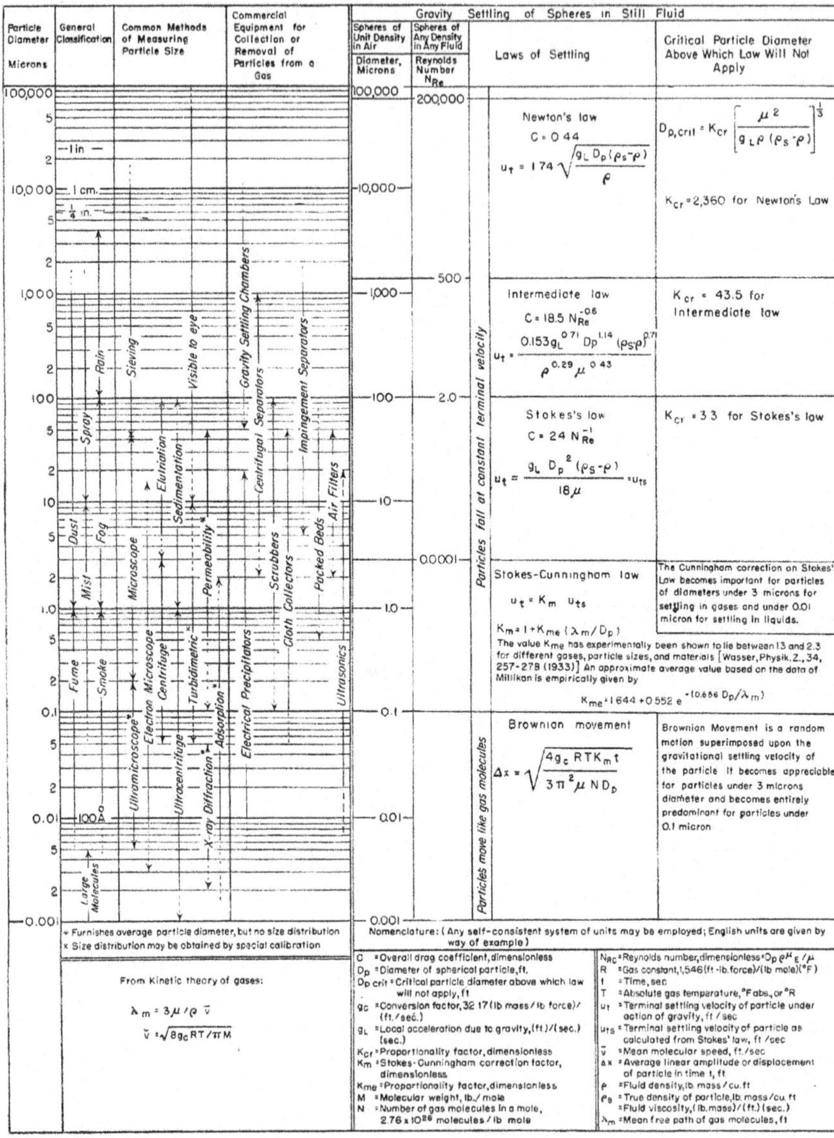

Fig. 113. Characteristics of gas dispersoids.

Figure 59 Particle Size—Characteristics of Gas Dispersoids[64]

Figure courtesy of McGraw-Hill Education

Chemical Engineers' Handbook, Third Edition, John H. Perry, Editor

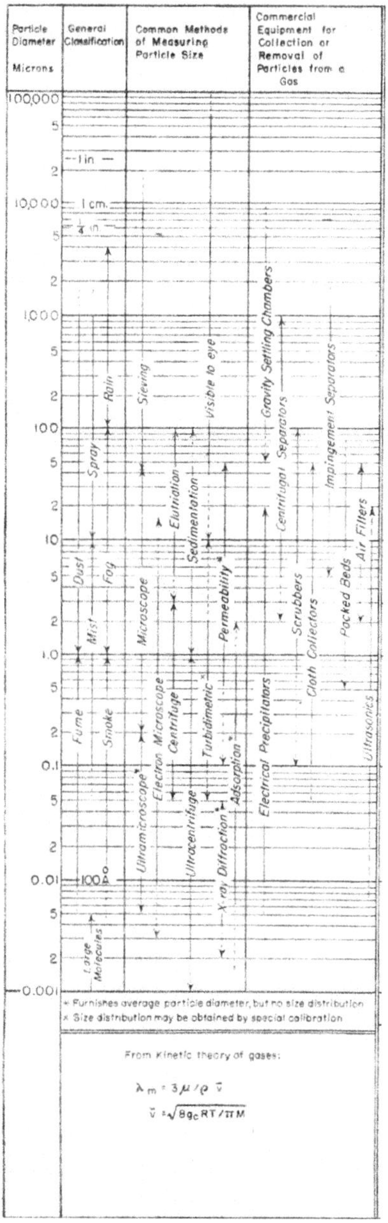

Figure 60 Types of Equipment to Remove Particles from Gas[64]

Figure courtesy of McGraw-Hill Education

***Chemical Engineers' Handbook,* Third Edition, John H. Perry, Editor**

increase the effective particle size to the point that "gravity settling chambers" or separators might be the best choice for removing those particulates from a system. Maintenance pigging is typically the most cost-effective way to remove agglomerated particles that have settled within pipelines.

References are readily available for providing an overview of separation technology and equipment.[65] However, as can be seen in Figures 59 and 60, each separation technology has a range within which they are effective. Although Figures 59 and 60 show centrifugal separators to be effective down to 2 μm, more recent literature from manufacturers may provide different limits. For example, one manufacturer cites the efficiency of their cyclone separators only down to 8 μm.[37] Hence, use those figures as guides, and consult with the equipment manufacturers for the most current specifications.

Select Separation Equipment to Match Particle Size of Black Powder

- Obtain samples of black powder from the system, as is, whether wet or dry
- Measure the particle size
- Select, modify, or replace process separation equipment, based on the typical range of particle sizes to be removed

11.3.1 Estimating Total Volume of Corrosion Products on Interior Surface of a Pipeline

The total volume of corrosion products or depositions within a pipeline is commonly estimated, based on a straightforward geometrical calculation. Calculate the circumference for the interior of the pipeline. Then multiply the circumference by the length of the pipeline. The product of those two numbers represents the interior surface area. Then determine the volume of those corrosion products or depositions based on the product (multiplication) of the internal surface area and the height of those corrosion products or deposition, as measured above the base metal within the pipeline.

Such an approach works well when estimating the buildup of iron oxide and iron sulfide corrosion products within gas pipelines, and can also be used to estimate the volume of paraffin deposited on pipe wall if the pipeline is transporting a paraffinic crude oil.

The sample calculations presented in this section are based on an assumed uniform formation of black powder on the interior of a pipeline transporting natural gas. However, if a pipeline has two or three phase flows, corrosion products or depositions may be more concentrated on the lower half of the pipeline. For that case, use knowledge regarding the proportion for each of the two or three phases, and the potential buildup of the corrosion products or depositions. Calculate the potentially affected cross sectional areas and the total volumes from the length of the pipeline segment.

Consider as an example a 30-in pipeline (76.2 cm), having a nominal wall thickness of 0.344 in (0.874 cm). Assume the pipe length is 10 mi (16.09 km) and the deposition is 1 mil (25.4 μm), where 1000 mil = 1 in. The mass or "weight" of the corrosion products removed can be estimated, based on the density (mass per volume) of the material removed:

The volume of solids within a pipeline is calculated by multiplying the surface area, by the height of those depositions.

The mass of those solids or corrosion products is calculated by multiplying the estimated volume by the density of the corrosion products.

Sample Calculation for Determining Weight of Corrosion Products

- The **interior circumference** would be **$C = \pi D$** = 3.1416 × (30 – 0.344 – 0.344) in for the two wall thicknesses. Thus, C_i = 92.086 in.
  Metric C_i = 3.1416 × (76.2 – 0.874 – 0.874) cm = 233.898 cm
- The **length (L)** of the pipeline segment is 10 mi (16.09 km)
  L = 10 mi × (5280 ft/mi) = 10 mi × (5280 ft/mi) × (12 in/ft)
  Thus L = 633,600 in
  Metric L = 16.09 km = 16,090 m = 1,609,000 cm
- The **height (H)** of the corrosion product or deposition is 1 mil = 0.001 in
  Metric H = 25.4 μm = 0.00254 cm
- **Total Volume** = $C_i \times L \times H$ = (92.086 in) × (633,600 in) × (0.001 in) = 58,345.690 in^3

Convert the Imperial Volume into cubic feet, where 1 ft^3 = 1728 in^3
(This conversion is made because Imperial Densities are in pounds per cubic feet (lb/ft^3.)
Thus, **Total Volume** of corrosion product = (58,345.690 in^3)/(1728 in^3/ft^3) ≈ 33.765 ft^3
Metric $C \times L \times H$ = (233.898 cm) × (1,609,000 cm) × (0.00254 cm) ≈ 955,908.4 cm^3

- Density of Iron Oxide ≈ 5.2 g/cm^3 ≈ 325 lb/ft^3 Metric and Imperial units, respectively
  Densities of Fe_3O_4 varies from 5.1–5.2 g/cm^3 ≈ (318–325 lb/ft^3)
  Density of FeS ≈ 4.84 g/cm^3 ≈ (302 lb/ft^3)
- Hence for this example: the Weight of Iron Oxide Corrosion Product
  Imperial (325 lb/ft^3) × (33.765 ft^3) ≈ 10,974 lb
  Metric ≈ (5.2 g/cm^3) × (955,908.4 cm^3) ≈4,971 kg

11.3.2 Flow Modeling to Estimate Deposition Within Pipelines

Computer flow modeling can also be used to estimate the volume of any deposits within a pipeline. Such modeling would be based on pressure drops throughout the pipeline, as well as flow rates along the pipeline. However, the modeling is beyond the scope of this book, and the reader is encouraged to consult with flow modeling specialists. The accuracy of their modeling is typically improved or refined by comparing the results from a subsequent cleaning pig run to the predicted results.

11.3.3 Geometry Pigs to Measure Depth of Deposits Within Pipelines

Geometry pigs are specially instrumented pigs designed to pass through pipelines to identify the location and severity of unknown restrictions such as dents, changes in the internal diameter of pipes, or perhaps misalignments in the welding of two pipeline segments. These pigs are not designed for cleaning pipelines, but instead to record measurements along the internal diameter through cantilevered feeler gauges, which ride along the interior diameter of the pipeline, covering all clock orientations. The feeler gauges have been used to measure the height of scale or black powder depositions within a pipeline. Consult with the maintenance pigging service providers to learn the options most suitable for any particular pipeline.

12 Modeling Transportation or Accumulation of Black Powder

Chapters 11 and 12 discuss different aspects related to particle size and the transportation of black powder particles through pipelines. Chapter 11 illustrated sample collection directly from an operating pipeline, such that particle sizes can be determined. Particle size is a very significant factor in modeling the transportation of those particles through pipelines, and will be presented in this chapter.

This chapter will reference the original paper used (see Sect. 12.2 and Ref. 69) for modeling the transportation of solids in horizontal pipelines, and several follow-up studies,[19-20] which focused on solid particle movement in oil and gas pipelines. Graphics from those studies are reproduced here, so that the reader can use the graphics and determine whether black powder may be accumulating at specific locations within their systems, based on factors such as flow rate or velocity, pipe diameter, and particle size. Sweep velocities are also presented.

12.1 Surface Roughness Increases from Corrosion or Black Powder Depositions

The interior surface of new pipelines is relatively smooth. However, over time they will become rougher, whether from corrosion, mechanical damage, such as from abrasion caused by the passage of solids, like sand. Alternately, the pipelines may appear rougher due to the accumulation of solids, such as corrosion products and black powder. From an operational perspective, the concerns are two-fold. First, the mechanical integrity of pipelines must be maintained. Second, when the surface roughness increases, there will be an increased pressure drop through the pipeline, which results in the need for additional compression to transport product through the pipeline. This will result in an increased cost in fuel for the compressors that are necessary to transport the product through the pipeline. Besides for possibly plugging pipeline appurtenances, such as used for instrumentation or product sampling, black powder can also abrade base metals at pipeline bends. As such, accumulations of black powder should be avoided as much as practical, if for no other reason than to lower the expenses for compressor fuel.

Table 9 provides the relative surface roughness in microinches (μin) and micrometers (μm) for new pipes (both coated and uncoated), as well as pipes that have been in service for 10 years, and a corroded segment of pipe. The relative surface roughness for a pipeline segment containing black powder can be anywhere in the range from < 250–18,000 μin (< 6.35–457 μm) depending on the volume of localized accumulations of black powder.

Table 9 Relative Surface Roughness[19]

Coating Status	Roughness (Imperial)	Roughness (Metric)
Internally coated (New)	< 250 μin	< 6.35 μm
Standard noncoated (New)	≤ 1800 μin	< 45.7 μm
Noncoated after 10 years (average)	~ 4,200 μin	~ 106.7 μm
Potential for corroded pipe (internal)	~ 18,000 μin	~ 457.2 μm

Where 1000 μin = 1.0 mil (1/1000 of an inch) = 25.4 μm.

Table 9 shows that a pipeline becomes a little bit rougher after 10 years of service. However, a corroded segment of pipe could become up to tenfold rougher, compared to that of new pipe. That means greater fuel expenses due to horsepower requirements for transporting the product, plus additional expenses for maintenance pigging and inspections to ensure system integrity is maintained. If there are accumulations of black powder within a pipeline, those depositions will increase the effective surface roughness.

Figure 61 compares the roughness of the interior surface of a pipeline and the reduction in revenue with the increased deposition of black powder. This graph illustrates the relative surface roughness in the 30 in (76.2 cm) gas transmission pipeline (a) before maintenance pigging, (b) immediately after pigging, (c) three months after pigging, and (d) five months after pigging. The increased expenses and reduction in revenue are directly related to the accumulation of black powder within the pipeline. Hence, this shows the true value of a maintenance pigging program for removing the accumulations of black powder.

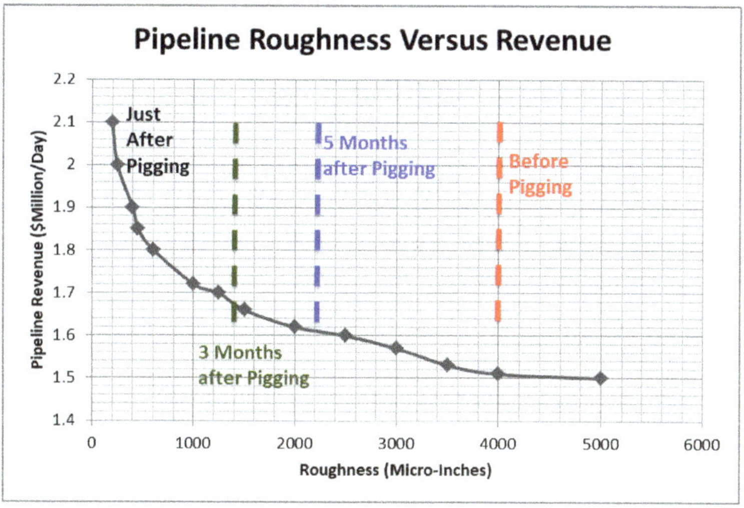

Figure 61 Increased Surface Roughness from Black Powder Deposition[67]

This particular graph was developed based on data for only one major gas transmission pipeline. The same relative decrease in revenue and increase in fuel expenses for compressors would be expected for other gas transmission pipelines, depending on flow rates and the quantity of black powder.

Although not specific to black powder depositions, another paper showed maintenance pigging reduced segment roughness from an average of 4700 μin down to 95 μin (119.4 μm down to 2.4 μm), based on six cases. They reported the pigging increased capacity by 28%, the reduction in pressure dropped by 39%, and there was a reduction in horsepower requirements of 23%, which translated to a fuel savings.[68]

12.2 Modeling the Movement of Black Powder

This section uses results from flow modeling of solids within a horizontal gas transmission pipeline to show that the volume of such depositions can be estimated. The velocity within pipelines directly correlates to the transportation or deposition of particles in those pipelines.

This section will present results from modeling studies for the transportation of solid particles, and in particular, focusing on particle size and the velocities necessary to transport those particles through a pipeline. The results from such modeling studies are used to illustrate the velocities necessary to transport particulates through pipelines, ensuring particulates do not accumulate. The graphics, which are presented later in Section 12.2, illustrate the significance of variables, such as pipeline diameter, flow rates, line pressure, etc.

The modeling for assessing the transportation of black powder in gas pipelines[19–20] was based on an original modeling study of flow through horizontal pipelines that M. Wicks published in 1971.[69] Please refer to the original Wicks publication and the follow-up modeling studies for in-depth discussions related to the mathematics used in the modeling.[20, 69] The following subsections provide an overview graphic results from that modeling.

12.2.1 Factors Used in the Modeling for Transportation of Particles Through Pipelines

Consider factors used in the modeling for flow or transportation of particles through a pipeline. Each particle within a pipeline is affected by gravity, buoyancy, lift, and drag, which varies with flow conditions. Gravity and buoyancy predominate at low flow rates, and particles can settle out to form a layer of sediment. The sediment layer can continue to build, until the cross sectional area of the pipeline has been reduced to the point when the localized velocity above the depositions is sufficient to push or transport particles along the direction of flow. Rounded particles will roll, while long, thin particles may slide or tumble. Rough or irregularly shaped particles may bounce along the pipeline, while flake-like particles might move like leaves in the wind. At higher velocities, the particles will follow the flow of the fluid, with some segregation due to gravity. However, at even higher velocities, the gravitational influences become negligible compared to the viscous forces.

The original Wicks publication was based on nominally dry particulates, and did not consider the presence of corrosion inhibitors, compressor oil, lubricants, glycols, or other chemicals, which can be found in gas transmission or export pipelines.[69] Although it is recognized that such liquid components serve to combine fine particles together to form larger particulates, it was not practical to model such attractive forces due to their complexity. (However, an empirical factor is reported later, which correlates results from theoretical dry pipelines to pipelines systems containing small volumes of liquid components that enable the dry particles to agglomerate into larger particles.)

12.2.2 Production Rates to Ensure Transportation of Particles Through Pipelines

The recent study of black powder movement within pipelines yielded Figure 62, which illustrates the velocities required to transport a 1 µm black powder particle within natural gas having a 0.63 sp gr., where 1000 µm = 1 mm.[19–20] The required velocities will vary, depending on the pipeline diameter and the pipeline pressure. Note that pipeline pressure affects the gas density, and hence affects the buoyancy and drag on the particles.

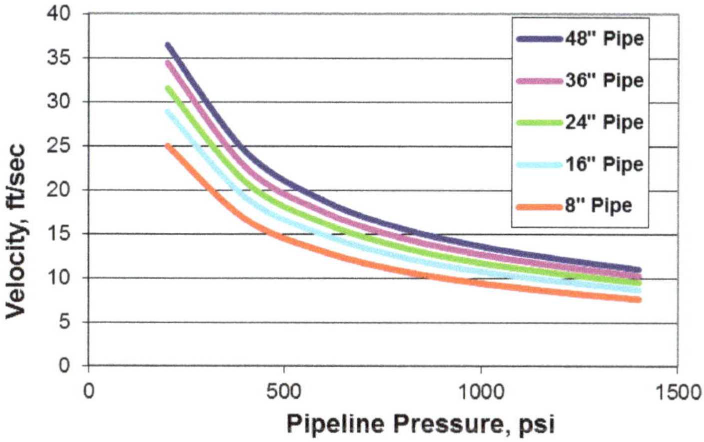

Figure 62 Velocity Required to Transport a 1 µm Iron Sulfide Particle in 0.63 sp gr. Gas[19-20]
(Graph from original publication; note 1 psi = 6.895 kPa and 1 ft/s = 0.3048 m/s)

12.2.3 Modeling Flow Rates for Transportation of Black Powder in Pipelines

Figure 63 presents much the same information as Figure 62. However, flow rates are used, rather than pipeline velocities, for five common pipeline diameters.

The flow rate required to sweep dry, solid particles through any particular pipeline diameter will decrease as pressure decreases. Thus, once a solid particle begins to move, it will continue to move until reaching a point where flow conditions change, such as at pipeline connections.

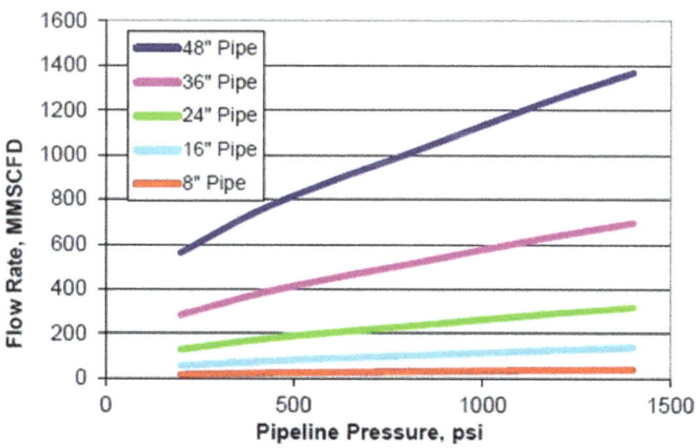

Figure 63 Flow Rates for Transporting 1µm Iron Sulfide Particle in 0.63 sp gr. Gas[19-20]
(Graph from original publication; Note 1 psi = 6.895 kPa and 1 MMcfd = 2.832 × 10^4 m^3/d)

12.2.4 Modeling Particle Size and Velocities

Numerous studies reported that black powder particles are brittle and will shear as they are transported along a pipeline. The modeling shows the smaller or finer particle size will reduce the velocities necessary to sweep the particulates down a pipeline. However, it was noted that the effects are not significant until particle size is less than 0.1μm. This is shown in Figure 64, for 0.63 sp gr gas at 68 bar (1000 psi = 6895 kPa).

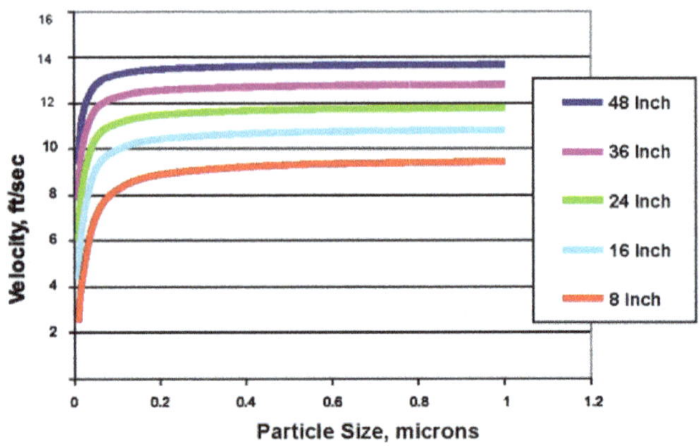

Figure 64 Velocities Required to Transport Iron Sulfide vs. Particle Size[19-20]
(Graph from original publication; pressure is 1000 psi = 6895 kPa; 1 ft/s = 0.3048 m/s)

Figure 64 shows it is possible to have the operational challenge of fine particulates being swept through a pipeline system, while larger or potentially conglomerated particulates may lay dormant. However, practically speaking, the black powder on the bottom of a pipeline is well mixed in particle size. While a small portion of the small, fine-sized powder may flow initially, most of those particles will be covered up by larger particles. Thus, expect very little movement until the velocities are such that the larger sized particles are moved. The net effect is that a pipeline, which has a significant volume of black powder, might not show any evidence of black powder until the velocities reach a critical velocity, at which virtually all the black powder and particles begins moving through the pipeline.[70]

As an example, one gas pipeline, which had black powder particles ranging from submicron to over 100 μm, operated for many years at velocities less than the entrainment or critical velocities necessary to transport the largest particles. When the flow was increased above that critical velocity, the black powder came in at a rate of approximately six tons (5443 kg) per day.[70]

12.2.5 Modeling Velocities for Transportation of Different Components of Black Powder

The relative velocities required to transport the various components of black powder in a 24 in (61 cm) pipeline at 60 °F (16 °C) are illustrated in Figure 65. Note that this is theoretical, because black powder found within pipeline systems is comprised of multiple components–not one single component. Figure 65 shows that the type and shape of the particles directly relates to the velocities required for transportation of those particles.

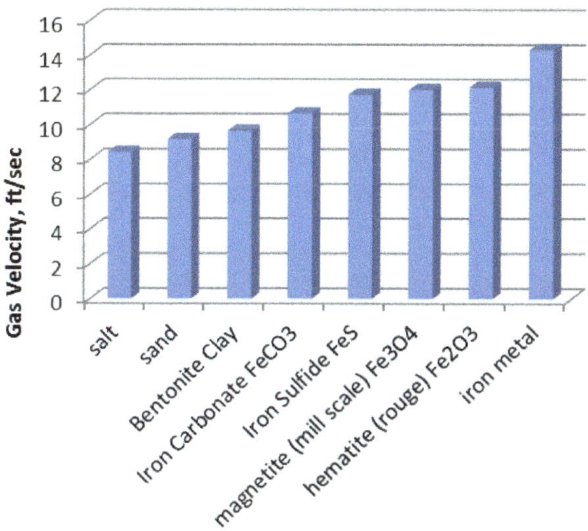

Figure 65 Velocities to Transport Various Particles[19-20]
(Graph from original publication; pressure is 1000 psi = 6895 kPa; 1 ft/s = 0.3048 m/s)

Figure 65 shows that theoretically some components of black powder may be transported more quickly than other components. However, the practical reality is that the black powder within any particular pipeline is typically a mixture of different sized particles, along with sand and liquids, such as lubricating oils or produced or condensed water. The fluids will typically provide adhesive forces between particles. Thus, anticipate the minimal movement of individual components, until the velocities have been increased to the point that the largest particles can be transported.[70]

12.3 Modeling Bed Height and Sweep Velocities

Perhaps one of the best visualizations for the concept of bed height and sweep velocity came when investigating the root cause for the generation and accumulation of black powder reported within a processing facility in Colorado. Specifically, a short pipeline segment between two flanges was removed, allowing for direct visual observation to the interior of the piping and the ability to grab a fresh sample of that powder for chemical analysis. Figure 66 is a close-up of a sample of black powder being collected from the interior of a pipe within a processing facility. The powder, which was pitch black, and filled the bottom half of the horizontal pipeline segment, was very difficult to photograph due to the contrast with the adjacent, bright sunlight illuminated section of piping.

It was possible to measure the height of the black powder depositions on the interior of the pipeline once the pipeline spool piece was removed to allow direct observation. Although the general flow rates were known for each pipeline segment, the bed height at that location effectively reduced the cross sectional area, causing a corresponding increase in the local flow velocity. The powder accumulated whenever the velocities were lower than those of the sweep velocities required for transporting the black powder. The particles are transported when the velocities through the reduced cross sectional area are above the sweep velocity required for transportation of the particles.

The black powder filled approximately half the height of this pipeline's interior diameter.

Figure 66 Samples May be Collected from Interior of Pipe When Accessible
Although difficult to photograph, the bed height of black powder deposition extended approximately half the internal diameter of the pipe (red arrow.)

After collecting samples for analysis, a room air fan was placed on the opposite end of this pipeline to blow the fine particulates from this particular pipeline segment. Figure 67 illustrates the fine black particles being blown out of this particular segment.

Figure 67 Fine Black Particle Being Blown out of Pipe Segment with Air Fan
The flow is from upper half of pipeline cross section, above the particle bed.

The velocities within pipelines can be used to estimate the height for the beds of particles, which may build up. Knowledge of the volume of particles or material within the pipelines can be useful when planning maintenance pigging or cleaning activities. The volume estimates can also be used as a potential measure to quantify the effectiveness of any remediation measures, which is commonly called a "metric," and should not be confused with the system of units for measurements.

Figure 68 illustrates the bed height versus pipeline velocity for a horizontal segment of 24-in (61 cm) pipe at 1000 psi (6895 kPa) and 60 °F (16 °C), assuming "dry" black powder. This illustrates a sweep velocity of approximately 13.1 ft/s (4.0 m/s). Thus, if velocities are higher, no accumulations would be expected. However, if velocities are lower, black powder particulates would accumulate until the bed height builds up and restricts the cross sectional area of the pipe segment, causing the velocities within the unrestricted portion of the pipeline's cross sectional area (above the black powder bed) to equal or exceed the sweep velocity.

Figure 68 shows that if the velocity was only 10 ft/s (3.05 m/s), the bed height could be 5 in (12.7 cm) from the bottom of the pipeline, and at approximately 4.4 ft/s (1.34 m/s), the pipeline could be half filled with solid depositions. The value of such modeling is that it makes it possible to estimate the volume of black powder that could be retained within a system, and should be considered when planning maintenance pigging operations.

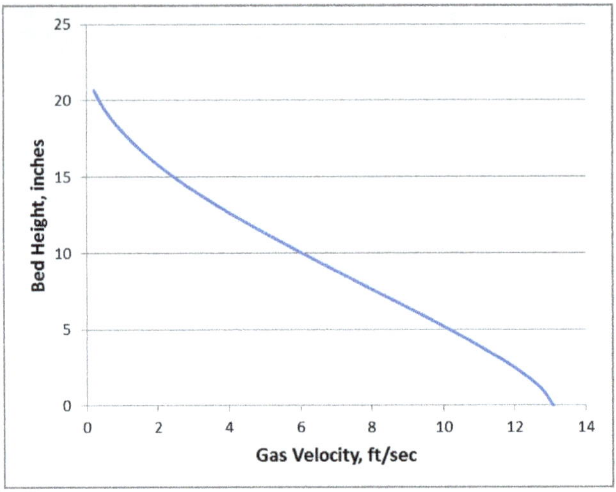

Figure 68 Bed Height vs. Gas Velocity for 24-in Pipeline[19-20]
1000 psi (6895 kPa), 60 °F (16 °C)
(Graph from original publication; 1 in = 2.54 cm; 1 ft/s = 0.3048 m/s)

Gas velocities and bed height can be used to estimate the volume of black powder in pipes.

12.3.1 Sweep Velocities for Inclined Pipeline Segments

The velocities required to move particulates uphill in a 24-in (61 cm) diameter pipeline were also modeled. This required a comparison between the flow requirements for a horizontal segment, versus those for a segment having a 30° inclination (Figure 69). As should be expected, slightly higher velocities were required to sweep the inclined pipeline, although the differences became very small when the black powder beds extended virtually all the way across the pipeline.

Modeling for Determining Entrainment Velocities

- Figures 58–61 and 64–65 present results from modeling entrainment velocities for 1 μm iron sulfide black powder particles
- Those figures can be compared to parameters from other systems to assess potential for transport or accumulations of such "dry" particles
- Particulates may accumulate when gas velocity is below sweep velocities
- The volume of black powder within a system can be estimated, based on known gas velocities and measured height of black powder beds

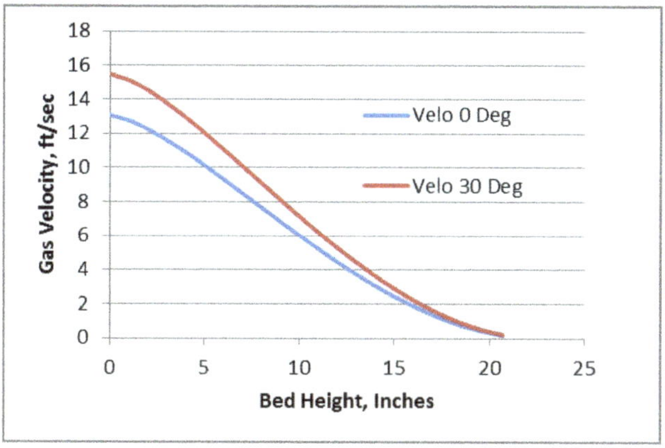

Figure 69 Bed Height vs. Gas Velocity for Horizontal and 30° Inclined (24-in) Pipe[19-20]
Note the horizontal and vertical axes are switched vs. Figure 68.
(Graph from original publication; 1 in = 2.54 cm; 1 ft/s = 0.3048 m/s)

12.4 Transporting Oily Black Powder

The discussions above were based on modeling flow and velocities, and assumed dry black powder, where there was no adhesion of particles. This simplification was necessary, due to the complexity for additional modeling of the adhesive forces associated with thin liquid films surrounding individual particles. The liquids could come from the presence of any free or condensed water or hydrocarbons, such as from glycol dehydrators or lubricating oil from pumps. The net effect of the presence of such liquids would be the agglomeration of smaller, fine particles into larger sized particles, which is attributed to the adhesive forces from the thin liquid films surrounding the individual particles.

Regardless of where any liquids are introduced, those liquids will elevate the adhesion between particles, and require higher velocities to break the adhesive forces and sweep the particles along the pipelines. Unfortunately, the modeling for such additional adhesive forces adds too much complexity to that modeling, and consequently, adjustments are made, based on comparisons to experimental data.[20]

Samples of black powder, which are collected from pipeline systems, commonly contain a mixture of water (produced or condensed), and hydrocarbons, such as condensed hydrocarbons, glycols from dehydration operations, and oils from compressors, which has been swept through the pipelines. Samples collected from operating systems typically appear oily, rather than being "bone dry," here meaning that there are no adhesive forces between particulates. Hence, a study was needed to generate a "correction" factor to provide the correlation between modeling results, which assumed nominally dry black powder, and oily black sample, commonly obtained from operating systems.

Accordingly, the flow rates or velocities necessary for transporting particles of black powder were determined for a pipeline, which happened to cross a river. In this field study, comparisons were made between the required velocities necessary to transport (a) nominally dry black powder, and (b) oily black powder. Based on those field studies, it was estimated that the velocity for moving oily black powder is approximately 2–2.5 times the velocities necessary for transporting dry dust black powder.[20]

Modeling Velocities for Transporting Oily vs. Dry Black Powder[20]

Based on field studies, velocities to transport oily black powder is 2–2.5 times the velocities required to transport dry black powder. This has been attributed to the adhesive forces associated with wet, oily, or glycol-wetted particles.

13 Overview of Chemical Treatments as Related to Black Powder

Numerous chemical products are used in the production of crude oil and natural gas. The types of products include corrosion inhibitors, demulsifiers or emulsion breakers, paraffin inhibitors, scale inhibitors, hydrate inhibitors, drag reducing agents, biocides, oxygen scavengers, hydrogen sulfide scavengers, chemical chelants, and many others. Of course, each of those types of products can be broken into multiple categories. There are both oxidizing and non-oxidizing biocides, and their functionalities are different. For example, aldehydes, such as glutraldehyde attack cell walls, while amines and quaternary amines target cell membranes, and chlorine attacks the cytoplasm. Likewise, paraffin inhibitors can be designed for flow assurance, or they can be crystal modifiers, meaning those paraffin inhibitors react with the nuclei of paraffin to change the surface chemistry, such that the paraffin does not conglomerate into waxy plugs. The topics are so numerous that a discussion would fill many books. Accordingly, this chapter will highlight just a few chemicals and their proper application for controlling conditions that could lead to the formation of iron sulfide and black powder.

13.1 Controlling Potential Corrosion from Residual Hydrotest Water

Whenever a new pipeline is constructed, it will typically undergo a pressure test to demonstrate the integrity of the pipe, as well as each of the welds and pipeline appurtenances. For safety reasons, the pressure tests used to document the integrity are typically conducted using pressurized water, and such tests are commonly called hydrostatic pressure tests. Although well intentioned, whenever water is used for conducting the pressure test, that water can pose a long-term threat to the integrity of the pipeline if it has not been properly treated to remove microorganisms and (to a much lesser extent) any dissolved oxygen. This section provides a discussion regarding the threat posed and how that threat can be properly managed through a combination of maintenance pigging, the (quick) removal of the water used in the pressure testing, and chemical treatments. The discussion is relevant to this book, as the primary threat posed is from sulfate reducing bacteria (SRB) within the test water and the hydrogen sulfide (H_2S) gas they generate. The H_2S, which the SRBs generate as part of their metabolic processes, is highly acidic and will etch and dissolve the base metal of the steel piping, forming iron sulfides, i.e., black powder.

First, recall Figures 43 and 44 (in Chapter 8), which illustrate microbiologically influenced corrosion (MIC) that penetrated the wall of a new pipeline before it went into service.[56] That failure was attributed to the presence of SRBs. Note the smooth round pits, and pits within pits observed on the interior surface of the pipeline segment at the location of the through-wall penetration in Figure 44. Also note the dark, black appearance of the base metal away from the pits depicted in the photograph. The black coloration is consistent with that of iron sulfide. (The interior of the pipeline sample had been cleaned at the location of the pinhole leak to make it easier to observe the base metal.)

Indeed, failures can occur in pipelines, which have never gone into service, if a portion or "heal" of the test water remains within the pipeline for an extended period of time, and the source water contained SRBs. The remainder of this section (Section 13.1) presents points to consider and address for properly managing the threats posed whenever conducting hydrostatic pressure tests.

13.1.1 Corrosion Control During Construction and Pressure Testing

Whenever any new pipeline is constructed, it is typically subjected to a hydrostatic pressure test to demonstrate the integrity of the pipeline segments and each of the welds. The pressure test is completed before the pipeline goes into service. If the pipeline is lengthy, it is typically fabricated in a number of different sections or segments. Thus, the integrity of each section should be tested and confirmed through a series of hydrostatic pressure tests–one for each segment. The construction project plan should include site-specific procedures for conducting the hydrostatic pressure testing, including identification of the sources of the water to be used in the testing. At the very least, those test procedures should call for filtration of the test water to at least 50 μm.

13.1.2 Corrosion Control Through the Source of Test Water: Water Quality is Critical

Water used for hydrostatic pressure tests are typically obtained from local sources, such as lakes, ponds, or streams. That is a practice designed to reduce expenses associated with transporting quality test water, such as potable water from municipalities, to the test site. (The primary concern associated with the use of water from lakes, ponds, or streams relates to possible microbiological corrosion that could result from the presence of any sulfate reducing bacteria (SRB) or acid reducing bacteria (APB) within any pools of water remaining within the pipeline following dehydration operations. Chemical treatment of the test water would mean the test water could not be returned to the original source following the testing.) Following completion of each of the hydrostatic pressure tests, those waters are typically returned to the original source. This is generally considered an acceptable practice, as long as the new pipeline never operated, nor contained any hydrocarbons or other chemicals. (The specific requirements associated with the return of test water are typically delineated in environmental discharge permits obtained through governmental agencies.)

Whenever new pipelines are constructed across remote lands, the costs for transporting large volumes of potable water are considered too expensive, and consequently water is often obtained from rivers or remote lakes or ponds. Unfortunately, those waters often contain high populations of SRBs and APBs, which could pose a threat to the integrity of the new pipeline if left within the pipeline for extended periods of time (typically > 28–30 days). Although it is generally recommended to collect and analyze the microbiological populations of test waters, such testing via serial dilution studies could require 14–28 days, respectively, for assessing SRB and APB populations. Construction crews would have typically completed their jobs, including dehydrating the pipelines and moved on to the next project before results would become available from serial dilution studies. Hence, it is incumbent upon the pipeline owner/operator to ensure preservative measures (chemical treatments) are implemented if the dwell or residency time of any small volumes of the test water remaining within the systems would be there for extended periods of time.

13.1.3 Corrosion Control: Beware of Extended Residency Time for Test Water

The maximum desirable time between completion of a hydrostatic pressure test and startup of pipeline operations is less than 28–30 days, based in industry practices.[56] Whenever layup times are expected to exceed 28–30 days, preservative measures, such as the application of combination oxygen scavenger, corrosion inhibitor, and biocide products, are recommended.[56]

13.1.4 Corrosion Control Through Pipeline Dehydration

Pipeline dehydrations are at best 90–95% efficient in removing any residual hydrostatic pressure test water. Hence, the pipeline owner/operator should be aware of the potential for a small volume of test water remaining within the pipeline, even after dehydration operations. There may be a continuing threat to the integrity of the pipeline posed by the presence of SRBs or APBs within any remaining pools of the test water. Accordingly, chemical treatments are recommended for preserving the pipeline from potential internal corrosion if residency times may exceed 28–30 days. (The pipeline previously depicted in Figures 43 and 44 was dehydrated, but not chemically treated, and a pinhole leak formed during the year-long period of time, prior to startup.)

Note also that the dehydration operations (maintenance pigs to displace and remove the water within the pipeline being tested) are typically conducted without the use of chemicals. Since the methodology described here is directed towards the testing of new pipelines, which have never been in service, there should be no problems, nor concerns, related to the return of that test water to the original source, as no chemicals were added to the source water. Nevertheless, coordinate with Environmental Discharge specialists to ensure proper discharge permits have been obtained.

13.1.5 Corrosion Control via Chemical Treatments for Preservation

If a pipeline will not begin operations for at least 28–30 days following completion of the hydrostatic pressure test, consider applying a chemical treatment to preserve the pipeline from internal corrosion, such as MIC caused by the presence of SRBs or any APBs within the test water. One advantage for conducting the preservative measure after removing the bulk of the test water from the pipeline is that a much smaller volume of chemical would be required for an effective treatment. As opposed to treating the entire volume of water within the pipeline, the preservative batch treatment would be based only on chemically treating the small volume of the remaining heal within the pipeline, which effectively reduces chemical expenses.

The preservative treatment would be based on displacing a small volume of potable water, containing a biocide to treat an assumed 10% remaining heal or volume from the original test water.[56] The concentration of biocide would be 500 ppm active. This is based on the total volume of water, i.e., the volume of the potable water to be used in the biocide treatment, plus the assumed volume of water remaining within the pipeline after completion of the pipeline dehydration operations (conservatively assumed to be 10% of the volume of the water used in the hydrostatic pressure test). Samples from this biocide treatment would be collected at the pig receiver to confirm that the entire length of the pipeline had been treated and as such, preserved for an extended layup. This confirmation would be based on analysis of the liquid samples collected at the pig receiver to confirm adequate concentration of the chemical treatment throughout the length of the pipeline. Such an approach would also make H_2S generation and iron sulfide formation unlikely.

13.2 H_2S Scavengers

H_2S scavengers are chemicals that chemically react with and remove H_2S from a process stream. Gas sweetening, or the removal of H_2S, can be through the use of iron sponge scrubbers, which form iron sulfide, or chemically, such as through the use of amine-based products.[71]

Large gas processing facilities typically use amine-based regenerative H_2S scavengers. That means that after the H_2S is absorbed at one step in the process, the H_2S is extracted in the second step to form elemental sulfur, after which the amine is returned to scavenge additional H_2S. (This is through a modified Claus Process, a discussion of which is beyond the scope of this book.)

The most common types of amines used in major gas processing systems, which regenerate their H_2S scavengers, include monoethanolamine (MEA), diethanolamine (DEA), N-methyldiethanolamine (MDEA), diisopropylamine (DIPA), and diglycolamine (DGA). Most new amine gas sweetening processes are based on the use of MDEA, which typically only absorbs H_2S, because systems using MEA, DEA, and DGA would also absorb other gasses, such as CO_2.

When the systems are much smaller and economics does not support the regeneration of the scavenger, nonregenerative scavengers are used to remove the H_2S. Perhaps the most widely used of the nonregenerative scavengers is triazine, which is a combination of the 1,2,3-triazine, 1,2,4-triazine, and the 1,3,5-triazine chemical structures. Aldehydes, such as glyoxal, will scavenge the H_2S, but at a much slower rate. Metal carboxylates and chelates have also been used for treating drilling fluids, as well as contaminated water and oil streams.[71] However, the workhorse for the industry is triazine.

Section 2.4.1 had a brief discussion regarding the storage of transmission quality natural gas within storage reservoirs or caverns. Such storage is an excellent tool which pipeline owners/operators use for ensuring capabilities to provide large volumes of natural gas in response to changing demands, such as those caused by changing weather conditions. However, any moisture or water within those reservoirs or storage caverns will make it possible for resident SRBs to flourish and release H_2S as part of their metabolism.

Although the natural gas being injected into such storage caverns may meet pipeline specifications, the gas retrieved from the storage facility could potentially deviate from the transmission specifications, particularly for the parts per million of the H_2S and the moisture content. Gas storage facilities should have instrumentation in place to monitor the retrieved gas and detect deviations from the specifications. That would make it possible for pipeline owner/operators or those personnel operating gas storage facilities to apply chemical treatments and bring retrieved gas back into specifications for return to the transmission pipeline.

For example, if flow from an individual storage well or cavern has sufficiently elevated H_2S that it would take the blended flow within a transmission pipeline out of the H_2S specifications, consider the injection of a water soluble hydrogen scavenger, such as triazine, into the flow from the individual reservoir or cavern. The triazine treatment would remove the H_2S from that gas. The treatment would be through the application of a water-soluble H_2S scavenger, and any excess of that scavenger (and the small volume of water necessary for that treatment) could be removed at the dehydration unit, before the retrieved gas is returned to the gas transmission pipeline. The removal of the H_2S and moisture will also significantly reduce the likelihood that any iron sulfide or black powder could form within the gas transmission pipeline.

13.3 Chelants to Remove Iron Sulfide

Chelants are specialty chemical products used to clean metal surfaces and remove corrosion products. The word comes from the Greek word for claw. A chelant will selectively "pinch" and bind with a specific metallic ion, such as calcium or iron. Then, once bound, it will remain in solution, but have much lower reactivity. One of the most commonly known chelants is ethylenediaminetetraacetic acid commonly called "EDTA."

Consider chemical cleanings, such as the boilers used in the power industries. Cleaning was historically undertaken to remove hard water depositions, such as calcium deposits. Originally, solutions of hydrochloric acid, which contained corrosion inhibitors, were used for chemical cleaning of the boiler tubes in the power industries. However, for safety and other concerns, chelants are now commonly used for chemical cleaning.

When attempting to clean iron sulfide or black powder depositions from a pipeline, perhaps the most important parameter is the ability to react quickly, while handling the materials safely. In contrast, when a boiler or heat exchanger is to be cleaned, the cleaning solution will be circulated through the system, with perhaps multiple passes. The chemical treatment plans are thus developed, based on site-specific operations.

The cleaning or removing of black powder corrosion products from a pipeline presents a much greater logistical challenge due to the total volume of the chelant solution needed for each cleaning run through a pipeline, noting that multiple pass recirculation is not practical.

The only practical methodology to apply such a cleaning solution to the interior of a pipeline would be by placing the cleaning solution either in front of one maintenance cleaning pig that forms a seal with the pipeline, or between two pigs that maintains a seal while passing through the pipeline. The drive or force for the maintenance pigging operations would have to be sufficiently strong, such that the cleaning solution maintains contact with the entire circumference of the interior of the pipeline.

The chemical cleaning product needed for cleaning pipelines to remove iron oxide and iron sulfide corrosion products must be fast acting, and capable of transporting the corrosion products to the terminus of the pipeline for removal either within the downstream separators or the pig receivers.

Tetrakis(hydroxymethyl)phosphonium sulfate, better known as "THPS," is such a product. THPS has the functionality to serve both as a biocide and as a chemical chelant. It was used as a biocide to re-establish control of microbiological populations within a subsea pipeline that was breached during the passage of a hurricane, as previously discussed in Section 6.3. However, by virtue of limited options in applying the product, the as-applied concentrations of that product were such that the product also functioned as a chelant, cleaning the pipeline by removing the iron oxide and iron sulfide corrosion products. The use of THPS is further discussed in Chapter 14.

14 Case Study: Pipeline Cleaning Using Solution of THPS

This chapter starts with an overview for the properties of THPS. Next, the discussion from Section 6.3 will be revisited for the point of illustrating the role that THPS had in reestablishing control over the microbiological populations of sulfate reducing bacteria (SRBs) and removing iron sulfides. Then, another example of using THPS for cleaning a pipeline will follow.

14.1 What is THPS?

THPS is an acronym for tetrakis(hydroxymethyl)phosphonium sulfate. Per the Product Information sheet, the structure of THPS is illustrated in Figure 70.

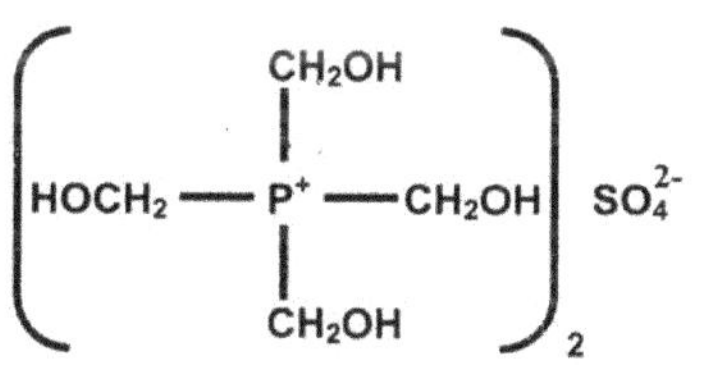

Figure 70 Tetrakis(hydroxymethyl)phosphonium Sulfate (THPS)[72]

THPS is a rather unique chemical, which was initially formulated for its ability to suppress fires. Subsequently, it was found to have excellent properties as (a) a chemical chelant, (b) an antifoulant, and as (c) a biocide. "Chelant," as previously described in Section 13.3, means that THPS reacts with and removes iron sulfide from pipelines. THPS is also a registered biocide, and as such, has the ability to kill or control microbiological populations. It also has the capability of scavenging oxygen, provided 20 ppm active THPS are used to treat each 1 ppm of oxygen.

Sidebar: Defining Active Concentration

The term "active" refers to the activity or concentration of a product. Thus, if the solution of THPS had 25% strength, it would be necessary to apply the product at four times (4×) the volume to achieve the desired strength for a treatment. For example, if 100 ppm of a chemical is being added to a mixture, and it is only 25% strength, then it would be necessary to add that chemical at 400 ppm to achieve the desired 100 ppm active.

THPS is typically applied at 50–200 ppm when used as a biocide, such as during treatments of vessels or tanks, since there would be a longer contact or treatment time. Concentrations are typically closer to 1000 ppm for applications, such as inside pipelines, when there would be a brief or limited contact time for killing the bacteria. However, when used as a chelant for quickly dissolving iron sulfide deposits, the concentration of THPS would be in the range of 20–25%. The application of THPS into the pipeline (which was described in Section 6.3 and will be discussed further in the next section, Section 14.2), was intended primarily for controlling the population of SRBs which flourished within the pipeline following the pipeline breach, when that subsea pipeline filled with seawater. Once repaired and the bulk of the seawater had been removed, the THPS was injected into the pipeline from offshore platforms, and was pushed through the pipeline by pigs to ensure proper delivery and application of the product. There were limited options for applying the chemical product at offshore platforms. The treatment also had to be based on the estimated volumes of seawater still remaining within the pipeline, the estimated volumes of produced water entering the pipeline system from downstream production, and the large volume of product necessary to treat the near 100 mile (161 km) pipeline to the onshore facilities. As such, the concentrations of the biocide treatments applied for treating those affected offshore pipelines, was also sufficient for the THPS to function as a chelant.

Figure 71 depicts THPS once it chelates with iron sulfide.

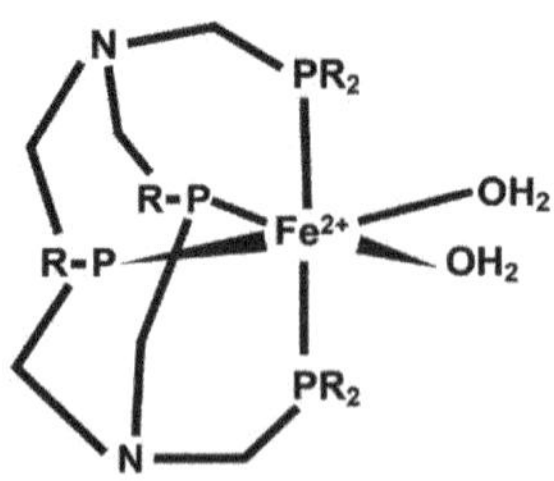

Figure 71 THPS Following Chelation Complex with Iron[72]

THPS has become a widely used biocide and anti-foulant cleaning agent for the petrochemical industry. It is high performance, fast acting, noncorrosive, biodegradable, and a low aquatic toxicity microbiocide. THPS can be denatured with a 3% solution of hydrogen peroxide, generally making it possible to discharge into the environment.[8] However, it will still be necessary to review and follow local regulations regarding the discharge of any solutions that contained THPS or other products.

14.1.1 What is THPC?

More recently, tetrakis(hydroxymethyl)phosphonium chloride (THPC), a variation of the THPS product, has become commercially available. Features include low toxicity, effectiveness at low dosages, and rapid breakdown in the environment. The structure of THPC is nearly identical to that of Figure 70, except that chloride replaces the sulfate, and only one THP cation is required to match the valence of the chloride.

Presently, only a limited number of publications are available that discuss the performance of THPC. Although preliminary, the feedback has been very encouraging, and the performance has been reported to be very similar to that of THPS. Consequently, it would not be unreasonable to anticipate upcoming publications to further document the performance of THPC as a potential biocide and chelant.

14.2 THPS Applied at Concentration Necessary for Chelation Functionality

The use of THPS was initially presented in Section 6.3, which described the remedial chemical treatments applied, following repairs of a subsea pipeline.[52-53] That pipeline had filled with seawater when the lateral was ripped from the trunkline during the passage of a hurricane. The chemical treatment program consisted of a combination of biocides, corrosion inhibitors, and maintenance pigging. The biocide selected for eradicating the sulfate reducing bacteria, which flourished within the pipeline, was THPS. It was clearly needed, as evidenced by the 90^+ ppm H_2S found in the seawater samples collected from the pipeline, following the repairs, while the trunkline was being dehydrated.

The discussion in that section included a graphic, which depicted results from analysis of fluid samples collected onshore, coinciding with multiple maintenance cleaning pig runs for the first nine months after the pipeline was returned to service. That data included measurements for the concentration of iron, the concentration of H_2S dissolved within any produced or seawater still being removed from that offshore gathering system, the concentration of THPS, noting multiple treatments, the pH, and stain tube measurements for the concentration of H_2S within the gas produced offshore. Note that the THPS treatment for this offshore pipeline was applied, based on THPS's properties as a biocide, and not for its complementary function as a chemical chelant. The treatments were designed to eradicate any remaining populations of SRBs, and the large treatment volumes were necessary due to the limited options for applying chemical treatments to that offshore, subsea pipeline, and the near 100 mile (161 km) path to the onshore processing facilities.

Long after the offshore gathering pipeline had been returned to service, a connection was made to a new subsea pipeline, which linked to a different gathering system. (Such additional connections provide flexibility to operations personnel in bringing product to the onshore facilities for processing.) After the new pipeline connection was made, the small cutout from the original pipeline (where the connection was made), became available for visual and metallurgical examination. That small cutout, which is also called a "pipeline coupon," was taken to the onshore facility and was subsequently examined. The metallurgical examinations confirmed that pipeline coupon was in excellent shape, only having a light surface oxide and no evidence for the presence of iron sulfides.

Originally, it was assumed that preexisting iron oxide corrosion products may have prevented the 90^+ ppm H_2S from forming iron sulfides. However, in retrospect it is more likely the high volume and concentration of the application of THPS for eradicating any microbiological populations also enabled the THPS to function as a chelant. Thus the THPS functioned as a chelant and evidently removed any iron sulfide or black powder corrosion products from the pipeline, explaining why iron sulfide deposits were not observed on the pipeline coupon, once it became available for observation.

Specifically, the large volume batch of commercially available THPS was applied at approximately 20–25% strength at the start of the offshore pipeline (from an offshore platform) and slowly pushed through the pipeline, using a scraper disc pig, which maintained a tight seal with the pipe wall. The concentrations of the THPS would ensure a quick kill of any microorganisms remaining within the pipeline, and the concentration of the THPS was also sufficient for the chelation functionality to remove any iron sulfide or iron oxide corrosion products. That explains why deposits were not observed on the pipeline connection coupon.

Also note that during the treatment applications, the speed of the pig was controlled by the offshore production. The contact time, which is the amount of time the chemical treatment would be in direct contact with any particular point along the pipeline, ensured the THPS to be an effective biocide, even if diluted by heals or volumes of residual seawater that may have remained within the pipeline.

14.3 Cleaning an East Texas Pipeline Using THPS

This section provides a discussion regarding another field application of THPS for cleaning a pipeline that contained iron sulfide black powder.

14.3.1 Background and Concerns from Increased Volume of Black Powder

One producer in East Texas experienced a significant increase in the volume of ferrous corrosion products removed from their system through maintenance pigging, after switching chemical service providers and the specific product, which was injected downhole to protect their wet gas wells.

This particular producer has two wells in a gas reservoir, which together produce approximately 4–10 MMscfd (1.33×10^5 – 2.83×10^5 m^3/d) of natural gas, plus condensates and produced water. The produced gas and liquids are transported to a gas processing facility through 4-in (10.2 cm), 6-in (15.2 cm), and 8-in (20.3 cm) pipelines, which connect to a 10 in (25.4 cm) pipeline, which can be pigged. The facility processes approximately 20 MMscfd (5.66×10^5 m^3/d) of natural gas, plus approximately 6,000 gal (22,700 L) of condensate per week, and handles approximately 5,000 gal (18,900 L) of produced water daily. Approximately 10 gal (37.85 L) of demulsifier, otherwise known as "emulsion breaker," were injected daily into the process piping just upstream of 12 liquid separator tanks to ensure the effective separation of the produced water from the hydrocarbon condensates.

Although analysis of the gas at the wellhead showed the gas contained essentially no carbon dioxide (CO_2) and no hydrogen sulfide (H_2S), nevertheless, SRBs were found within the produced water. Those SRBs will react with and consume any sulfates present within the water in the reservoir to generate H_2S. The gas and liquids are produced at 150 °F (65.6 °C), which is ideal temperature for microbiological populations to flourish. Thus, H_2S generation and reactions to form iron sulfide are probable, if the microbiological populations of the SRBs are not kept in check. Otherwise, there could be issues related to the presence of black powder.

Operations personnel observed large volumes of black powder approximately one month after switching the chemical product service provider and the chemical product, which was injected downhole to control corrosion. Operations personnel also installed a new compressor just upstream of the production facility. This was primarily to maintain production rates. However, wet black powder clogged the 5 μm pores of a "witches' hat filter," which is a metallic, cone-shaped, mechanical filter installed within a pipeline to catch any particulates and protect downstream compressors. (Similarly, the conical strainer depicted in Figure 27 is such an inline filter, which had to be cleaned out every 2 hours (h), due to the rapid buildup of particulates and the resultant high differential pressure across those filters.)

The root cause for the increased volume of black powder the East Texas producer experienced could potentially be associated with the switch of chemical products, or perhaps related to the change in process equipment, such as from the new compressor. Regardless, the goal was cleaning the system and minimizing future formation of black powder.

A filter coalescer, which is capable of collecting particles as small as 0.7 μm was considered as a possible solution. The plan was to install it just upstream of the compressor. However, operations and corrosion control professionals realized that high differential pressures, such as observed across the witches' hat filter, would starve the flow into the compressor, causing it to shut down. (If the gas flow rate from the wells was reduced, the flow would bypass the compressors to keep the field wells online and producing.) Hence, a chemical cleaning approach was sought for removing the black powder.

Fresh samples of the black powder were collected from the witches' hat. The samples were magnetic, and x-ray analyses confirmed the powder was wet iron sulfide. Thus, knowing the black powder was primarily iron sulfide, the recommendation was made to clean the pipeline and facilities using THPS.

14.3.2 Chemical Treating for Removing Black Powder Using THPS

The product selected for this particular cleaning application was a 75% active concentrate of THPS. However, it was not going to be used at that strength. Instead, the treatment solution to be used was a 47% concentration of the 75% strength concentrate THPS product. Thus, the actual strength of the THPS used for this particular cleaning operation was approximately 35% active concentration of THPS in water.

Sidebar: Considerations When Transporting Concentrated Products

It is a common practice for the chemical product suppliers to ship concentrated products to specific client locations. Then, either a representative of the chemical product provider or the pipeline or processing company would dilute those products to the desired strengths or concentrations for the applications. Potable water is used for most dilutions. This practice of shipping concentrated products helps reduce overall costs, since the transportation costs associated with shipping the diluent (water) is eliminated

Sometimes low ambient temperature may be a concern. Consider, for example the potential problems that could result if a chemical product is chilled to the point that some components of concentrated products may precipitate or fall out of solution. That could compromise the performance of the product, as has happened to corrosion inhibitor products delivered to oilfield operations above the Arctic Circle. Consult with the chemical product technical representatives before adopting the use of concentrated products. There may be other options, such as changes in the solvent packages used when formulating those products

The field-blended 35% strength THPS was then injected into the pipelines from the two well sites at a rate of 24.5 gal/d (92.7 L/d) for three days. The treatment rate was based on production rate of approximately 10 MMscfd (2.83×10^5 m^3/d) natural gas and 2500 gal (9463 L) of produced water per well site. The black water, which had been observed at the gas plant, cleared up within three days.

A separate field separator and associated pipeline also became plugged with the wet black powder. Again, 35% active strength of THPS was applied for cleaning that field separator, using a 110 gal (416 L) batch, and both the separator and associated piping were successfully cleaned.

As operations returned to normal, the chemicals used for the downhole treatments were reviewed. The original product, which had been applied downhole, and the replacement product was not identified. However, once operations personnel and the corrosion control specialists recognized that microbiological populations were rising, a new replacement product was selected, based on having the ability to control corrosion, as well as secondary functionality as a biocide or biostat. Note that biostats are products that keep the microbiological populations under control, but do not necessarily kill the microorganisms.

The maintenance pigging of the well lines, which had been conducted annually, was switched to weekly when the witches' hat filters were clogging up every couple of hours. However, once the lines had been cleaned using THPS and the new downhole treatment program had been implemented, the maintenance pigs were rescheduled for once every 120 days.

15 Potassium Permanganate

Potassium permanganate ($KMnO_4$) is a strong oxidizing agent and provides an intense purple-pink color in water. It has been used to disinfect drinking water, as well as to remove iron and H_2S from well water.

Solutions of $KMnO_4$ have also been used as the first of a two-step process in chemical cleaning of the deposits that build up on the tube sheets of the steam generators used in the nuclear power industry. Chemical chelants, such as ethylenediaminetetraacetic (EDTA) acid, would constitute the second step for a chemical cleaning treatment.

The major chemical service providers, who support crude oil and natural gas gathering and production, are not offering this chemical, even as a first step oxidizer to enhance a second step use of chelants. The rationale is that other products are either more effective, or are available at a lower cost.

16 Case Study: Cleaning Pipeline Lateral Using Solution of Oxidizer and Surfactant

This case study addresses the cleaning of a pipeline lateral in West Virginia, USA. It is a fairly new pipeline in a new gas gathering system that is still under development. As such, there really was not much operational, nor corrosion monitoring data available to suggest concerns. However, operations personnel initially identified a concern when significant localized burn or melting was observed on a foam pig, which had been used to clean a pipeline lateral. The damage to this foam pig, which was illustrated in Figure 1 and is duplicated here in Figure 72, is consistent with what could be expected when pyrophoric iron sulfide comes into contact with oxygen, once the pig receiver was opened. The pyrophoric iron sulfide, which was removed from the pipeline during the pigging run, apparently smoldered and then burst into flames, releasing heat and melting the foam. Section 2.2.2, which addressed safe handling of maintenance pigs, included photographs that illustrated hosing down and cleaning the pigs in the receiver and containment pit to remove any residual black powder residing on the pig.

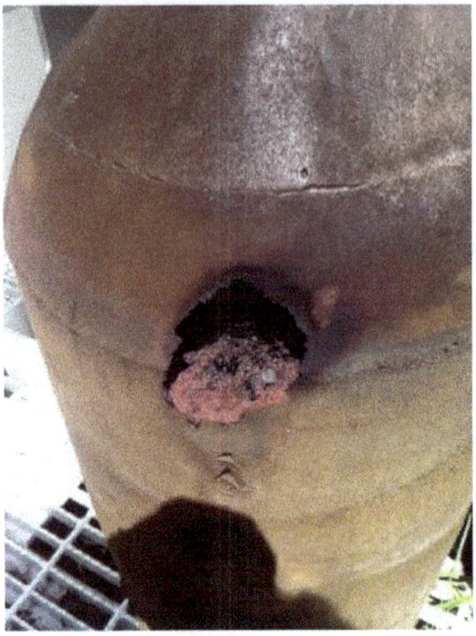

Figure 72 Foam Pig Damaged by Apparent Pyrophoric Iron Sulfide
Localized melting of the foam pig attributed to pyrophoric iron sulfide.

16.1 Description of Conditions and Need to Clean

This particular pipeline lateral is 12-in (30.5 cm) in diameter, and is 2.53-mi (4.07 km) long. It operates at 1,000 psig (6895 kPa), and flows at a rate of 5 MMscfd (14.16 x 10^4 m^3/d). The pipeline is pigged twice a week (Tuesday and Thursday), and a foam pig is used to displace the liquids down a pipeline. Although there is a receiver at the end of the pipeline, there was no collection tank at the receiver. Thus, the produced liquids are swept through a "blow-by" line and thereby diverted into the downstream pipeline, which was also a 12-in (30.5 cm) diameter pipeline. All liquids are separated at the processing facility.

16.2 Field Samples to Determine Concentration of H_2S and CO_2, and Moisture Content

The gas is sampled at the receiver for H_2S, CO_2, and moisture content. Typical readings were 2 ppm H_2S, 0.1 mole percent CO_2, and 3 lb/MMscf (48 mg/m^3) water vapor. (Note: countries using metric units typically report moisture content as a water dew point in degrees Celcius (°C), based on a set pressure, such as 40 bars, which is ≈ 4054 kPa). The gas velocity was estimated as 1.09 ft/s (0.33 m/s).

The process for assessing potential threats for internal corrosion is to first consider the partial pressure of CO_2 within this pipeline to determine whether corrosive conditions may be present. As noted above, the line pressure for this particular pipeline is 1000 psig (6895 kPa). Next, determine whether the moisture content of the gas may provide the electrolyte necessary for the corrosion mechanisms. Then check the velocities within the pipelines.

- The H_2S results would typically be considered "good," because the concentration of H_2S is approximately half of the commonly used pipeline specification for sales gas, which is ¼ grain/100 ft^3, or approximately 4 ppm.)
- The 0.1 mole percent (Mole%) CO_2 results would be considered "extremely low." Typically, the carbon dioxide content of produced gas is 0.1–4.0 Mole % for gathering systems and 0.1–2.0 Mole % for gas transmission pipelines. The potential threat posed by the presence of CO_2 is usually assessed, based on the partial pressure of CO_2, relative to the other components of the gas. These will be presented in the next section. However, in this example, the partial pressure of CO_2 indicates generally noncorrosive conditions.
- A 3 lb/MMscf (48 mg/m^3) moisture content is considered to be extremely dry. Pipelines are typically considered dry from the perspective of sales gas if the moisture content is 7 lb/MMscf (112 mg/m^3) or less. For offshore pipelines transporting natural gas through cold, deep water, the specifications are typically 2.5–3.0 lbs/MMscf (40–48 mg/m^3). That is to minimize the chances for any condensation and possible formation of hydrates or internal corrosion. The low moisture content within this particular pipeline suggests a reduced likelihood for the presence of condensed or free water, which would be necessary to support any corrosion mechanisms.
- The gas velocity of 1.09 ft/s (0.33 m/s) cited for this example is considered slow, as typical pigging is at 5–7 mi/h (8.0–11.3 km/h), or approximately 7.3–10.3 ft/s (2.2–3.1 m/s). Thus, if any liquids were present, there should be some evidence of those liquids within the debris or corrosion products, which the cleaning pigs move or displace into the pig receiver.

16.2.1 Use Partial Pressure of CO_2 to Identify Corrosive Conditions

The partial pressure of carbon dioxide (P_{CO2}) can be used to determine whether corrosive conditions may exist within a pipeline or process equipment.

Use Partial Pressure of CO_2 to Identify Potentially Corrosive Conditions

Total Pressure of a System is the sum of the partial pressures of the gasses present.
Partial Pressure CO_2 = P_{CO2} = Total Pressure × [(Mole % CO_2)/100%]

General Rule of Thumb:

- **P_{CO2} > 30 psi (207 kPa) indicates corrosive conditions probable**
- **P_{CO2} between 7 and 30 psi (48 and 207 kPa) indicates corrosive conditions may exist**
- **P_{CO2} < 7 psi (48 kPa) indicates generally noncorrosive conditions**

The above applies to vertical tubing.
Corrosion can occur on horizontal lines at partial pressures < 7 psi (48 kPa) where water collects (pools).

Consider a pipeline with **0.1 Mole % CO_2 at 1000 psig (6895 kPa):**
P_{CO2} = (1000) × (0.1%/100%) = 1 psig **(6.895 kPa)**
This suggests noncorrosive conditions.

Consider a pipeline with **1.0 Mole % CO_2 at 1000 psig (6895 kPa):**
P_{CO2} = (1000) × (1.0%/100%) = 10 psig **(68.95 kPa)**
This suggests corrosive conditions may exist.

For the case of 0.1 Mole % CO_2 content, corrosive conditions would NOT be expected.

16.2.2 Calculate Velocities from the Flow Rates

The velocity of the gas being transported can be calculated based on the known production rate, such as 5 MMscfd (1.42×10^5 m^3/d), the line pressure, and the internal diameter of the pipeline.

To obtain the velocity in feet per second, convert the days to seconds. Additionally, any volume measurement, such as 5 MMscf (1.42×10^5 m^3/d), is based on standard atmospheric pressure, namely 14.7 pounds per square inch absolute (psia) (101 kPa). The actual volume transported under pressure must be calculated using the pipeline's operating pressure, plus the atmospheric pressure. The actual volume of gas flowing under pressure is based on 1000 psig (6895 kPa) plus 14.7 psi (101 kPa) atmospheric pressure, or 1014.7 psia (6996 kPa). Use Charles's Gas Law to calculate the volumes flowing under pressure (psia) and the pipe's cross-sectional area to calculate the velocity.

Calculating Flow Rate through a Pipeline

Given: 5 MMscfd (1.42×10^5 m^3/d) = flow rate.

1 ft^3= 1728 in^3 (from 12 in/ft) (1 ft^3 = 0.02832 m^3)

1 day = 24 h = 24 × 60 min = 24 × 60 × 60 s = 86,400 s

From Charles's Gas Law $P_1V_1 = P_2V_2$

P_1 = Ambient Pressure = 14.7 psi

P_2 = 1000 psig +14.7 psi = 1014.7 psi (6996 kPa)

Cross-sectional area for interior of pipe = $\pi\ (ID/2)^2$.

For 12-in pipe, OD = 12.750 in (32.4 cm)

Let thickness of pipe wall = 0.410 in, hence, ID = 12.750 in – 0.410 in – 0.410 in = 11.930 in (30.3 cm)

Internal cross-sectional area = $3.14 \times (11.930/2)^2 = 3.14 \times (5.965) \times (5.965)$ in = 111.725 in^2 (720.85 cm^2)

Take Volume of Actual Gas Flow (under pressure). Then divide by cross-sectional area to provide the flow rate:

(5 $MMft^3$/d) × (1728 in^3/ft^3) [(14.7 psi)/(1000 + 14.7 psi)] × [1 d/24 h) × (1 h/60 min) × (1 min/60 s) = 1448.70 in^3/s (23,740 cm^3/s)

Flow Rate = (1448.70 in^3/s)/(111.725 in^2 cross-sectional area = 12.967 in/s.

Convert to feet per second (1 ft = 12 in)

Flow Rate = (12.967 in/s) × (1 ft/12 in) = 1.08 ft/s (32.9 cm/s ≈ 0.33 m/s)

--- -----

In the equations above,

OD = Outside diameter of Pipe ID = Inside diameter of Pipe

Charles Law: P_1 = Pressure 1; V_1 = Volume 1 P_2 = Pressure 2; V_2 = Volume 2

16.2.3 Insight from Knowing Velocities within a Pipeline

Just like high flow velocities suggest a potential concern for erosion or erosion-corrosion mechanism, low velocities should also be scrutinized. Low velocities may suggest that certain sized particles of black powder will not be transported through pipelines, but instead will accumulate up to the point of effectively reducing the cross-sectional area and increasing the velocity of the flow across the remaining portion of the pipeline's cross-sectional area. However, very low velocities may help in other ways, such as improving the effectiveness of chemical treatments.

Low flow velocities typically suggest that there would be good contact time for any chemical treatments applied to the interior of the pipeline via pigging. The key is to ensure the volume of the chemical treatment is sufficiently large, that as that volume of chemical is being pushed through the pipeline, there will be a minimum of **ten seconds** (10 s) contact with any individual point along the pipeline. The chemical treatment, applied within an aqueous solution, is sometimes referred to as a "chemical pill." The volume of that pill or aqueous chemical treatment is typically based on establishing a 1–2 mil (25.4–50.8 μm) film or layer of the chemical product on the interior surface of the pipeline, and adjusting it to account for any slippage as the pig traverses the full length of the pipeline. In addition, the volume may need to be adjusted, such that the chemical pill has a 10 s contact time with the pipe wall while being pushed through the pipeline.

16.3 First Cup Pig Run

The day following the foam pig run, a cup pig was run through the same pipeline, and approximately 2½ gal (9.5 L) of solids were received at the pig receiver. That is not entirely surprising in that foam pigs are designed to displace liquids through a pipeline, rather than push debris. The field notes clearly delineated the collection of solids, rather than a sludge or liquids. That observation is consistent with the gas having very low moisture content, such as 3 lb/MMscf (48 mg/m^3) water.

16.4 Field Tests Confirmed Presence of Iron Sulfide

A small sample or aliquot was collected and placed in a glass container, and those solids were wetted by a couple of drops of 15% hydrochloric acid. Lead acetate paper was placed across the opening of the glass container, and the test paper quickly turned brown-black. Additionally, there was a distinctive smell of rotten eggs. Next, a magnet was run across the exterior surface of the glass container, and the sample responded to the magnetic field. These field tests provided the distinct, positive indication for the presence of iron sulfide.

16.5 Recommendations for Chemical Cleaning Using a Surfactant and Antifoulant

A recommendation was made to chemically clean that particular pipeline, due to a combination of factors. These included (a) the observed melting of a portion of a foam cleaning pig, which could occur if patches of pyrophoric iron sulfide remained on the pig after it was removed from the receiver and was allowed to dry, and (b) analysis of the 2½ gal (9.5 L) of solids, which confirmed the presence of iron sulfide. A pipeline cleaning was necessary if for no other reason than safety. The product selected for this pipeline cleaning was a combination surfactant and an antifoulant.

16.6 Operations and Management Review

Shortly after the receipt of the 2½ gal (9.5 L) of dry black powder, the 12-in (30.5 cm) lateral pipeline was shut-in due to unrelated problems in the 12-in (30.5 cm) trunk line, which was immediately downstream. This provided an opportunity to review the primary safety concern, namely the smoldering of the black powder, which occurred within 30 seconds after exposure to air. Since the presence of pyrophoric iron sulfide was observed, one of the first steps would be to adopt the new industry practice and spray water into the pig receiver and onto the pig, as soon as it is removed from the pig receiver. If the black powder possibly contains pyrophoric iron sulfide, keep that material wet until it has been appropriately disposed.

The pipeline cleaning was based on the use of a chemical pill, which consisted of a mixture of 16 gal (60.6 L) of a chemical product: a combination surfactant and antifoulant, with 32 gal (121 L) of fresh water. Per the original plan, that blend would then be injected into the pipeline, just ahead of a cleaning pig. One of the first questions, however, was whether there was sufficient contact time for the chemical treatment to be effective. That leads to a quick calculation for verifying whether the contact time is sufficient, or whether the volume of the batch treatment might need to be increased, or the speed of the maintenance pigs applying the product might need to be adjusted.

Determining Contact Time for Chemical Application Via Pigging

- Let the cleaning pill volume be 16 gal (60.6 L) product + 32 gal (121 L) fresh water = 48 gal (182 L)
- There are 231 in^3/gal (3785 cm^3)
- Cleaning pill volume = (231 in^3/gal) × (48 gal) = 11,088 in^3 (181.7 L = 181,700 cm^3)
- Internal cross sectional area of 12-in (30.5 cm) OD pipeline = 111.725 in^2 (720.85 cm^2) (from above)
- Pill would occupy 11,088 in^3/111.725 in^2 = 99.24 in (252.07 cm) length within pipe
- Convert length of chemical pill to feet: 99.24 in/12 in = 8.27 ft (2.52 m)
- Flow rate per above = 5 MMscfd, equates to 1.08 ft/s (0.33 m/s) in a 12-in (30.5 cm) pipeline
- Contact time = 8.27 ft/1.08 ft/s = 7.66 s ≈ (2.52 m)/(0.33 m/s)= 7.64 s

This is slightly less than the generally recommended 10 second minimum contact time.

16.6.1 Ensuring Contact Time Sufficient for Effective Treatment

Accordingly, the corrosion control technicians adjusted the volume of the chemical pill to 20 gal (75.7 L) of the combination surfactant and antifouling product with 40 gal (151.4 L) of fresh water, making a 60 gal (227 L) pill size. That brought the contact time up to approximately **9.6 seconds.**

16.7 Subsequent Runs with Chemical Pill Ahead of Cleaning Pig

Over the course of the next two weeks, three two-disc sealing pigs and one foam-brush pig were launched on a Monday and Thursday pigging schedule. In each case, the chemical pill injected in front of the pig consisted of 20 gal (75.7 L) of the combination surfactant and antifoulant, with 40 gal (151.4 L) of fresh water. Table 10 summarizes the results from those runs, each at a gas flow rate of 5 MMscfd (1.42×10^5 m^3/d).

Table 10 Volume of Black Powder Removed per Run

Date[a]	Type of pig	H_2S ppm	CO_2	Water Vapor lb/MMscf (mg/m^3)	Volume and Character of Deposits
5/14/14	Foam	UNK[b]	UNK	3 (48)	Unknown Volume dry solids; suspected pyrophoric FeS burned and melted into pig
5/16/14	2 Disc Sealing–NO Pill[c]	2 at R[d]	UNK	3 (48)	2½ gal (9.5 L) Dry Black Solids–Smoldering when exposed to air/oxygen (within 30 s) FeS
6/16/14	2 Disc Sealing + Pill	2.5 at L[e] 2.0 at R[d]	0.1 at L 0.1 at R	16.8 (269) at L 13.1 ((210) at R	1/4 gal (0.95 L) moist black solids; could be formed like clay
6/19/14	2 Disc Sealing + Pill	2.2 at R	0.1 at R	14.2 (227) at R	1/8 gal (0.47 L) black fluid
6/23/14	2 Disc Sealing + Pill	2.6 at L 2.7 at R	0.2 at L 0.1 at R	17.9 (287) at L 20.7 (332) at R	1/16 gal (0.24 L) black sludge
6/26/14	Foam Brush + Pill	1.02 at L 2.45 at R	0.1 at L 0.1 at R	33.6 (538) at L 57.4 (920) at R	Dry Black Solids seen on part of bottom of jar. Field tests suggested 0.6 ppm H_2S in solids

a. Date is in month-day-year format
b. UNK = Unknown; Not Measured
c. Pill = 20 gal (75.7 L) surfactant and antifoulant product with 40 gal (151.4 L) fresh water ahead of pig
d. R = Receiver
e. L = Launcher

16.8 Final Observations on Use of Combination Surfactant and Antifoulant

The 12 in (30.5 cm) diameter pipeline lateral had historically been pigged regularly using a foam pig, which is designed to displace free liquids. However, foam pigs were not designed to displace solids, and would be expected to flex and "jump" over small accumulations of any solids within a pipeline. When the first cup pig was run through this lateral, it brought in approximately 2½ gal (9.5 L) of dry black powder. Subsequent runs brought in ¼ gal (0.95 L), 1/8 gal (0.47 L), 1/16 gal (0.24 L), and finally just a trace of solids. Thus, approximately five pig runs were required to remove the black powder solids from this particular pipeline.

After the passage of the first pig, which brought in 2½ (9.5 L) gallons of dry black powder, a combination product containing a chemical surfactant and an antifoulant, was injected just ahead of the pig. As should be expected from the injection of a chemical product blended into fresh water, the solids removed in the later runs were wet, and in the form of a slurry.

The initial analysis of the solids collected from the lateral's pig receiver confirmed that at least a portion of the solids were in the form of a pyrophoric iron sulfide, meaning that the black powder could spontaneously combust when exposed to air. That was confirmed, based on observation of the foam pig, which sustained local burn marks and melting of the foam. The use of the liquid pill containing the product, which had a combination of a surfactant and an antifoulant in fresh water, was a good choice for cleaning the pipeline, if for no other reason than keeping the solids removed wet.

17 Case Study: Cleaning Using a Surfactant, Enzyme, and Mild Oxidizer Product

This case study employs a specialty product, which had been used several times for cleaning tanks associated with crude oil and natural gas production in the Middle East. It summarizes one of the first attempts to employ this technology in North America. The process was apparently successful from the perspective of removing pyrophoric iron sulfide by converting the sulfide to elemental sulfur, which could easily be removed. However, one of the process steps called for temperatures to be elevated to approximately 1200 °F (650 °C) for one hour. Such elevated temperatures, which are an integral part of the cleaning process, may pose concerns related to the maximum temperatures or other restrictions for American Society of Mechanical Engineers (ASME) code tanks and pressure vessels. Accordingly, an ASME-certified tank and pressure vessel inspector should review and approve such elevated temperature cleaning procedures from the perspective of maintaining the certification and integrity of those tanks and vessels. Additionally, no information was available regarding the effect such temperatures may have on any gasket materials or instrumentation. Those considerations should also be addressed and documented when preparing for such cleaning operations.

17.1 Description of Natural Gas Liquid (NGL) Tanks to be Cleaned to Remove Pyrophoric Iron Sulfide

Samples of the Natural Gas Liquids (NGLs) were collected from the inlets to the NGL storage tanks in a West Virginia processing facility. Field and laboratory analyses of the samples of solids confirmed the presence of a pyrophoric form of iron sulfide. Thus, an initiative began for cleaning and removing that iron sulfide from those storage tanks, as it posed not only safety and corrosion concerns, but also concerns related to quality control for the NGLs, which are sold.

This facility has six 60,000 gal (227,100 L) horizontally oriented NGL storage tanks, each of which had all the NGL removed and the tank cleaned, such that threaded fittings could be replaced by welded fittings. Each tank was 11 ft (3.35 m) tall and 87 ft (26.52 m) long, and had one 18-in (45.7 cm) blinded manway at one end of the tank. The inlet and outlet drains were located on the bottom of the cylindrical tanks, and on the same end as the manway. One of those NGL tanks is illustrated in Figure 73.

Figure 73 Horizontally Oriented NGL Storage Tank

Figure 74 illustrates several of the NGL tanks and associated piping.

Figure 74 Multiple NGL Storage Tanks and Associated Piping
These tanks were to be cleaned to remove black powder/pyrophoric iron sulfide deposits. Once cleaned, the threaded fittings were replaced with welded fittings.

17.2 Cleaning Product–Aqueous Based Surfactant with an Enzyme and Mild Oxidizer

The product selected for this particular chemical cleaning was based on several criteria. These include the ability to eliminate the hazards presented by pyrophoric iron sulfide, as well as the ability to complete the cleaning operation in a short period of time, such as 24 hours. The product selected was described as a proprietary combination aqueous based surfactant, an enzyme, and a mild oxidizer. (The specific components were listed as proprietary.) The Materials Safety Data Sheet (MSDS) described the product as a liquid bacterial solution.

Per the product bulletin, "the chemistry is extremely unique in that it creates a light emulsion to de-oil equipment internals, but demulsifies when quiescent without the use of external emulsion breakers." The term "quiescent" is defined as "in a state or period of inactivity or dormancy." The product bulletin describes an aqueous based product, which will change the metal surfaces from oil or hydrocarbon wetted surface to water wetted, and the light emulsion, which is used to prepare the metal surfaces, will dissipate under near stagnant conditions. The product was also described as "biodegradable, non-exothermic, wastewater friendly, and compatible with all known metallurgy."

The product brochure described the application of the product as being through the circulation of either an aqueous solution or vapor phase. The vapor phase application calls for the product to be directly pumped into steam, which is then injected into the vessels or tanks. The product then contacts and penetrates the interior surfaces of the vessel or tank being cleaned. This application methodology was selected, because that approach would result in a lower volume of waste being generated. In addition, there would not be a need for recirculation pumps, and thus the total treatment time could be reduced.

17.3 Cleaning the NGL Tanks

Tanks 4, 5, and 6 were selected for cleaning. Final preparations, which verified all hoses connected, etc., started at 7:00 AM, and steam flow began at 11:00 AM. The chemical product injection began at 12:00 PM (noon) and continued until 4:30 PM. The key performance indicator for the cleaning process was the reduction of the percent lower explosion level (LEL) to the target 0% for each of the three tanks and the condensed water tank. Once the steam injection stopped, the temperatures quickly dropped. The manway blinds were then removed, and air movers were used for full ventilation, allowing for visual inspections the following morning. Data from the cleaning process is presented in Table 11 below.

Table 11 Chemical Cleaning of NGL Tanks to Remove Pyrophoric Iron Sulfide

Cleaning Tank 4	Time	% LEL	H_2S (ppm)	Product (ppm)	Temp °F (°C)
4	8:46	100	0	0	36 (2)
4	12:11	100	0	0	156 (69)
4	13:47	8	0	100	127 (53)
4	14:40	1	0	300	120 (49)
4	15:35	0	0	130	1200 (649)
4	16:00	0	0	140	1200 (649)
4	16:18	0	0	150	1200 (649)
Cleaning Tank 5	**Time**	**% LEL**	**H_2S (ppm)**	**Product (ppm)**	**Temp °F (°C)**
5	8:42	20	0	0	31 (-0.5)
5	12:09	100	0	0	86 (30)
5	13:45	55	0	0	155 (68)
5	14:38	5	0	200	150 (66)
5	15:30	0	0	160	1200 (649)
5	16:04	0	0	160	1200 (649)
5	16:15	0	0	165	1200 (649)
Cleaning Tank 6	**Time**	**% LEL**	**H_2S (ppm)**	**Product (ppm)**	**Temp °F (°C)**
6	8:35	0	0	0	33 (0.5)
6	12:07	100	0	0	88 (31)
6	13:40	14	0	0	170 (77)
6	14:33	0	0	175	200 (93)
6	15:20	0	0	175	1200 (649)
6	15:48	0	0	175	1200 (649)
Cleaning Frac Tank	**Time**	**% LEL**	**H_2S (ppm)**	**Product (ppm)**	**Temp °F (°C)**
Frac Tank	8:30	0	0	0	29 (-1.7)
Frac Tank	12:04	100	0	0	41 (5)
Frac Tank	13:37	100	0	0	78 (26)
Frac Tank	14:45	50	0	215	100 (38)
Frac Tank	15:15	30	0	220	1200 (649)
Frac Tank	15:45	0	0	210	1200 (649)
Frac Tank	16:05	0	0	215	1200 (649)
Frac Tank	16:20	0	0	212	1200 (649)

Samples obtained from Tanks No. 4, 5, and 6 were collected through 1 in sample ports on each tank. The samples collected from the Frac Tank were obtained at the manway cover for that tank.

A corrosion control technologist, who was not affiliated with the chemical supplier, collected the data. There were no indications for the presence of hydrogen sulfide throughout the test.

From the perspective of corrosion control, it is unusual to track the percent lower explosion limit (% LEL) as a key performance indicator for the chemical cleaning operation. Normally, the % LEL is checked primarily before personnel enter a tank or vessel to conduct an assessment of the interior surface and assess potential corrosion or the condition of a coating. However, in this case, the % LEL was used to monitor the chemical cleaning process and determine when the interior of a tank or vessel had been fully decontaminated to remove any remaining traces of the hydrocarbons. The measure for the effectiveness of the treatment came the following morning, once temperatures had cooled to ambient, and the vessels or tanks were opened to the atmosphere for visual inspection. If pyrophoric iron sulfide remained and was exposed to the atmosphere, it would smolder, and the vessels could not be safely entered for the visual inspection and removal of the elemental sulfur. The ability to safely enter the vessel, once it was cooled to ambient temperatures and open to the atmosphere (oxygen), demonstrates the apparent effectiveness of this particular chemical cleaning process in eliminating the threat posed by pyrophoric iron sulfide.

Following the treatment and cooldown, each of the tanks was entered, and the visual inspections confirmed any previously present iron sulfide had reacted to form elemental sulfur. That elemental sulfur was then removed from the NGL tanks by mechanical brushing, coupled with a water wash. The interior of each vessel was thoroughly dried to eliminate any pools of water before the vessels were returned to service.

Two things to note: First, new chemical cleaning products and application methodologies will undoubtedly be developed in the future. Whenever such new applications are being considered, make sure there are clear, unequivocal performance metrics or measures, which can be used to control the processes and to assess the effectiveness of the chemical application itself.

Second, this particular application ended with the vessels cooling down overnight. The high temperature application of steam just before the cooldown was apparently intended to neutralize the chemical product. If the enzyme was actually specialty bacteria, as suggested by the MSDS, the 1200 °F (649° C) steam treatment certainly would eradicate any bacteria or decompose any enzymes used within the treatment itself.

Whenever considering a maintenance or cleaning operation that requires the use of such high temperature steam, be sure to evaluate not only the processes used to remove corrosion products or deposition, but also the possible effects those temperatures may have on the system metallurgy, gasketing materials, and existing instrumentation. Another question to address in the pre-planning is whether the application of such high temperature steam could exceed any ASME code requirements for those affected vessels or tanks?

18 Strategy to Find Upstream Sources of SRBs, NORM, or Undesirable Gasses

Whenever results from sampling suggest one or more of the parameters, which are normally monitored, are outside of expectations, the first questions relates to the validity of that data point, and the source for that unexpected data point. This chapter presents a strategy for determining or identifying upstream sources for (a) unexpected excursions from the expected results, such as increased populations of sulfate reducing bacteria (SRBs) or acid producing bacteria (APBs), (b) elevated naturally occurring radioactive materials (NORM), or (c) perhaps increased concentrations of undesirable gasses, such as H_2S or CO_2.

18.1 Example: Sampling to Identifying a Well Needing Elevated Treatments

Consider for example, an offshore gathering system, which required very high treatment rates, which were expensive, and also posed logistical challenges for delivering the chemical treatments. Were the elevated treatments required for all wells flowing into the gathering system, or were the treatments just needed for select wells? A corrosion monitoring program can be used to determine where the treatments need to be applied, as well as determine the necessary treatment rates.

The impetus for this particular study was that the chemical treatment rates necessary for controlling corrosion on one particular pipeline in the Gulf of Mexico was approximately 16 times the generally recommended treatment rate (1 pt/MMscf ≈

16.7 L/MMm^3). The natural question was "Do the expenses for the purchase and delivery of the additional volumes of corrosion inhibitor exceed the commercial value from the production at that particular well?"

The cost-benefit analysis can proceed once the source for the elevated H_2S and CO_2 has been identified, the flow rates have been determined, and the necessary chemical treatment rates have been quantified. This chapter provides a strategy for systematically identifying upstream sources that may be contributing to downstream threats to system integrity.

18.2 Producer Sampled Each Source Flowing into the Production Line

The producer sampled production streams from each of the wells and determined that a specific well was introducing the bulk of the H_2S and CO_2 to the platform's total production of natural gas. The question the producer had to answer was whether there was more value or earned income from the natural gas produced by that individual well, versus the expense of the chemical treating (and processing) for that gas. Note that as the production or flow from that (or any particular) well drops due to normal well or reservoir depletion, a point will be reached, below which the value of the produced gas will not offset the expense for the additional corrosion inhibitors necessary to treat and protect the pipeline from internal corrosion.

The sampling program, which the producer initiated, was able to identify a specific well that had been introducing corrosive elements into the pipeline. Economics was then used to determine when to shut-in and ultimately plug and abandon that particular well, enabling lower treatment expenses while still maintaining the integrity of the subsea pipeline.

18.3 Strategy: Identify Concerns at Terminus, and Work Upstream to Find Sources

Figure 75 illustrates a gathering system feeding a slug catcher at the entrance to a processing facility. The well lines flow to one of the manifolds, and the lines from the manifolds flow to the slug catchers, which are at the entrance to the processing facilities. Assume that there is an indication of a problem or concern, based on analysis of samples collected at the slug catcher or end/terminus of that gathering system.

The indicator could be an elevated concentration of SRBs within any produced water arriving at the slug catcher for the facility. That would suggest the possibility of microbiologically influenced corrosion within the upstream pipelines, or that the SRBs could generate H_2S, which in turn could react with the pipe wall steel and initiate corrosion.

Alternatively, the indicator could be an elevated concentration of H_2S, relative to that expected, based on the most recent analyses of gas samples collected from individual wells. The volume of any black powder collected during maintenance pigging operations would also serve as an indicator.

Another indicator for a potential concern that merits investigation is any accumulation of debris, including corrosion products, black powder and formation fines, such as sand, within the slug catcher. This could suggest diminished capacity (functional volume), and possibly reduce the residency time within the vessel. Regardless of the specific concern, the approach for identifying the source would be the same.

Consider Figure 75. The first step to identifying unique sources for the concerns would be the collection of samples at the first points upstream from the slug catcher. Per Figure 75, the first place to look and sample would be at the manifolds A, B, and C. Then, if for example, elevated microbiological populations H_2S or CO_2 were found in samples collected at Manifold C, the next step would be to collect samples at Wells 7, 8, and 9. Then, following this same example, if elevated microbiological populations H_2S or CO_2 were found for Well 8 (but not for wells 7 and 9), it would be apparent that remedial measures or treatments would need to be initiated for treatments at Well 8. The simple strategy is that if unacceptable results are observed downstream, sampling would be initiated at the subsequent points upstream, until the source or sources have been identified. Once identified, treatments would be applied at the upstream sources, and the effectiveness of the remedial measures would be quantified, based on downstream measurements or sampling.

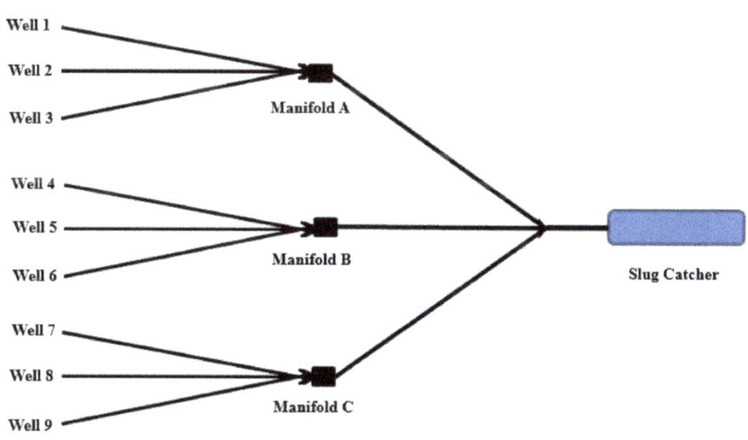

Figure 75 Gathering System and Collecting Samples to Find Source(s) of Contaminants

18.4 Strategy: Once Source is Identified, Treat and Monitor Effectiveness Downstream

Using the specific example cited above, if elevated SRB populations were found at Well 8, then remedial measures, such as biocide treatments, should be initiated at that point. The effectiveness of treatments would be based on monitoring and sample analyses collected at the downstream source, which in this example would be Manifold C.

> **If concerns are reported downstream, sample progressively upstream to find the source.**
>
> **Treat the source to address concerns, and monitor the effectiveness of treatments downstream.**

19 Establishing Metrics for a Black Powder Management Program

The effectiveness of any corrosion control program is measured through the use of some form of metrics. This could be through controlling the general and localized, or pitting corrosion rates to being less than 1 mpy (25.4 μm/y) general corrosion rate and less than 5 mpy (127 μm/y) pitting rate, per NACE SP-0775.[54] The effectiveness could also be measured by other factors, such as keeping microbiological populations at or below 100 organisms per milliliter. The specific metrics that are used should be selected by a joint team of operations and corrosion control personnel, based on the unique operating conditions for each operating location.

Following are two partial lists related to controlling corrosion and managing the accumulation of black powder within crude oil and natural gas production, processing, and transportation to their customers through a pipeline network. The first list (Section 19.1) identifies operating parameters, which can be measured and used for detecting changes in the operations and performance of gathering and processing system. The second (Section 19.2) is a list of possible metrics, which would be used when evaluating the performance or effectiveness of corrosion control programs, including the management of black powder.

19.1 Operational Parameters That May Provide an Understanding of Corrosion Control

The following are examples for the types of data that corrosion control professionals would like to be able to access as part of their overall responsibilities for maintaining system integrity. These parameters are not the metrics themselves, but instead the metrics are drawn upon them. The keys are the data trends over time, and keeping each parameter within previously defined minimum or maximum acceptable limits. If any reading or results from sample analyses suggest anything unusual or out of the normal trend, the underlying cause has to be investigated.

- Production rate for individual wells or pipelines, including volumes of gas and liquids
- Pressure
- Temperature
- Gas composition, including H_2S, CO_2, O_2, moisture content
- Liquid composition including population of SRBs and APBs, pH, dissolved solids, and dissolved H_2S
- Residual concentration of any corrosion inhibitor and any biocides
- Maintenance pigging, including the volume and composition of liquids and solids from each run
- Inspection results, including RT, UT, and data from ILI runs
- Corrosion monitoring, including coupons and electronic probes
- Chemical treatments, including type of product and treatment rate based on production

Such data collection parameters can be used for establishing metrics. For example, consider results from analysis of corrosion coupons at any one individual location. If the corrosion rate was 0.81 mpy (20.6 μm/y), then typically there would be no change in any associated corrosion inhibitor treatment rate. If the corrosion rate was 2.10 mpy (53.3 μm/y), then there would typically be no change in the treatment program until after data from the next coupon exposure was assessed and the trend was confirmed. However, if the corrosion rate was much higher, e.g., 8.92 mpy (226.6 μm/y), then an immediate change in the treatment program would generally be warranted. (However, following investigations, if there was an obvious, one-time cause for such a high reading, then changes may not be necessary.) That is how the program should work.

The metrics for the corrosion coupon monitoring program could then be based on the total number of corrosion coupons evaluated over the time period, the number of readings that suggest a need for changes in the treatment program, and the number of those changes in the treatment program that were implemented. That shows the program is working as designed.

When developing a program for managing the potential accumulation of black powder, consider tracking the concentration of H_2S entering the system, as well as the concentration of H_2S exiting the system. The difference in parts per million, multiplied by the gas flow rate in million standard cubic feet per day or cubic meters per day can be used to determine the volume of iron sulfide that may be generated. Is the rate for iron sulfide generation increasing, decreasing, or approximately the same? Also consider the change-out of any inline filter cartridges, which may be used to capture fine particulates upstream of critical process equipment or sales transmission pipelines. Are the change-outs or replacements of the filter cartridges, which are based on pressure differentials across the filter cartridges, more or less frequent, or approximately the same? If there is a significant change, look for changes in the operating conditions or volumes processed, which may explain any differences. Likewise, consider the volume of solids removed during each of the individual maintenance cleaning pig runs. Is the volume removed increasing, decreasing, or approximately the same? Consider analyzing such data, based on a daily accumulation rate, because the number of days between maintenance pigging runs may vary. In addition, determine whether the material removed is iron sulfide and if it is pyrophoric. As appropriate, consider the effect of treatments, such as the use of THPS, to chemically chelate and remove any iron sulfide, which may have formed on the interior of the pipeline due to the H_2S.

19.2 Develop Metrics to Document Effectiveness of Black Powder Management Program

Numerous and varied sets of metrics could be developed as part of an overall black powder management program. The following are a few examples, which can be trended over time:

- Volume of iron sulfide removed during each pigging run. Compare the volume removed to the results obtained shortly after any chemical treatment for removing such depositions.
- Track the frequency for the maintenance pigging runs, and do these cleaning runs following the planned schedule? What are the volumes of liquids and solids removed versus time? The key is to look for significant changes in the character or volume of the material removed, and to then document whether any changes in the pigging or maintenance program are warranted.
- Measure the moisture content of process stream, and determine whether each measurement of the moisture content is within the pipeline specifications. Near or continuous readings are best, because it would be less likely for an excursion to go undetected. Any condensed water can serve as the electrolyte for corrosion mechanisms.
- The concentration of H_2S can be measured within gas or dissolved within produced- or condensed water. Compare concentrations upstream and downstream to calculate the volume or weight of iron sulfide being formed. If upstream concentrations are lower than downstream concentrations, look for additional sources for the H_2S, such as a low spot where water can accumulate and sustain colonies of SRBs. Develop metrics based on the relative concentration of H_2S upstream and downstream on a pipeline and match the volume of iron sulfide generated to that removed by maintenance pigging.
- Track the upstream and downstream populations of SRBs. This is because SRBs are a potential source for H_2S, and as such the microbiological pop-

ulations should be monitored. Also track the dates and concentrations of biocides applied to correlate the effectiveness of chemical treatments to microbiological populations.

- Where practical, compare results from successive ILI runs for specific points. Flag significant differences, such as a 20% change in depth, for follow up investigations. Ensure each has been reviewed, and that a recommendation has been made for operations personnel, including if no change is warranted.
- Monitor via weight loss coupons or electrical resistance probes, where practical. Categorize results, based on acceptability of results, such as per NACE SP-0775.[54] The number of locations yielding higher corrosion rates would be one metric related to system integrity. Another metric for the effectiveness of the corrosion control program itself could be based on the number of locations having elevated corrosion rates, which require increases in the chemical treatment rates, based on quarterly corrosion monitoring (assessment every 90 days).
- Where practical, compare results from successive nondestructive testing (NDT) inspections, which would typically be radiographic testing (RT) or ultrasonic nondestructive testing (UT). Such points could be on impellers, extrados of elbows, or "tees." The number of locations for NDT inspections, versus the number of locations exhibiting potential erosion or erosion corrosion would be one metric related to the overall system integrity. A second metric that could be used to document the effectiveness of the program, would be the periodic review of the monitoring and inspection results for potential threats and documenting whether or not any changes to the corrosion mitigation programs were or were not implemented.

References

1. ANSI/NACE MR0175/ISO 15156-1:2015, "Petroleum, Petrochemical, and Natural Gas Industries—Materials for Use in H2S Containing Environments in Oil and Gas Production" (Houston, TX: NACE International, 2015).
2. Baldwin, R.M., "Black Powder in the Gas Industry—Sources, Characteristics and Treatment, Gas Machinery Research Council Report No. TA 97-4" (Dallas, TX: Gas Machinery Research Council, 1998).
3. Baldwin, R.M., "Black Powder Control Starts Locally, Works Back to the Source," *Pipeline & Gas Industry* 82, 4 (April 1999): pp. 81–87.
4. Baldwin, R.M., "Tariff Control of Constituents that Form Black Powder in Gas Transmission and Distribution Pipelines," Report to the Gas Machinery Research Council (Dallas, TX: Gas Machinery Research Council, 1999).
5. Baldwin, R.M., "Here are Procedures for Handling Persistent Black-Powder Contamination," *Oil & Gas Journal* 96, 43 (1998).
6. Baldwin, R.M., "Black Powder Problem Will Yield to Understanding, Planning," *Pipeline & Gas Industry* 82, 3 (March 1999): pp. 109–112.
7. Baldwin, R.M., "The Characteristics of Black Powder in Gas Pipelines and How to Combat the Problem," *Corrosion Prevention and Control* 47, 4 (2000): pp. 95–102.
8. Baldwin, R.M., P. Redmond, and R.H. Winters, "Contaminants in Sales Gas Pipelines," SwRI Project No. 18.14226 (Chantilly, VA: Pipeline Research Council International, 2009).
9. Baldwin, R., "Gas Pipelines Black Powder: Who Should Know or Care," Gas Machinery Conference (Dallas, TX: Gas Machinery Research Council, 2006).
10. Sherik, A.M., "Black Powder in Gas Transmission Pipelines," *Oil and Gas Pipelines Integrity and Safety Handbook*, R. Winston Revie, Ed. (Hoboken, NJ: John Wiley & Sons, 2015).

11. Sherik, A., S. Zaidi, E. Tuzan, and J. Perez, "Black Powder in Gas Transmission Systems," CORROSION 2008, paper no. 08415 (Houston, TX: NACE International, 2008).
12. Sherik, A., J. Perez, S. Jutaily, and A. AbdulHadi, "Composition, Sources, and Formation Mechanisms of Black Powder in Gas Transmission Pipelines," EUROCORR 2007 (London, United Kingdom: Institute of Materials, Mining and Minerals, 2007).
13. Sherik, A., "Black Powder: Study Examines Sources, Makeup, in Dry Gas Systems," *Oil and Gas Journal*, 106, 30 (2008): pp. 54–59.
14. Sherik, A., "Corrosion of Sales Gas Pipelines," EUROCORR 2012 (London, United Kingdom: Institute of Materials, Mining and Minerals, 2012).
15. Sherik, A.M., A.L. Lewis, A.H. Rasheed, and A.A. Jabran, "Effect of Triethylene Glycol on Corrosion of Carbon Steel in H_2S, CO_2, and O_2 Environments," CORROSION 2010, paper no. 10188 (Houston, Texas: NACE International, 2010).
16. Sherik, A.M., A. Rasheed, and A. Jabran, "Effect of Sales Gas Pipelines Black Powder Sludge on the Corrosion of Carbon Steel," CORROSION 2011, paper no. 11087 (Houston, TX: NACE International, 2011).
17. Sherik, A., and B. Davis, "Thermodynamic Analysis of Formation of Black Powder in Sales Gas Pipelines," CORROSION 2009, paper no. 09560 (Houston, TX: NACE International, 2009).
18. Smart, J.S., "Determining the Velocity Required to Keep Solids Moving in Pipelines," Pipeline Pigging and Integrity Management Conference (Houston, TX: Clarion Technical Conferences, 2007).
19. Smart, J.S., and R.H. Winters, "Black Powder Problems in Gas Pipelines," Pipeline Pigging and Integrity Management International Conference (Houston, TX: Clarion Technical Conferences, 2011).
20. Smart, J.S., "Black Powder Movement in Gas Pipelines," CORROSION 2011, paper no. 11089 (Houston, TX: NACE International, 2011).
21. Smart, J.S., "Black Powder in Gas Pipelines," American Gas Association 2017 TIMP Conference (Washington, DC: AGA, 2017).
22. Smith, S.N., and M.W. Joosten, "Corrosion of Carbon Steel by H_2S in CO_2 Containing Oilfield Environments," CORROSION 2006, paper no. 06115 (Houston, TX: NACE International, 2006).
23. Smith, S.N., and M.W. Joosten, "Corrosion of Carbon Steel by H_2S in CO_2 Containing Oilfield Environments—10 Year Update," CORROSION 2015, paper no. 5484 (Houston, TX: NACE International, 2015).
24. Little, B.J., and P. Wagner, "Microbiologically Influenced Corrosion," *Peabody's Control of Pipeline Corrosion*, 2nd ed., R.I. Bianchetti, Editor (Houston, TX: NACE International, 2001).
25. Pope, D.H., "State of the Art Report on Monitoring, Prevention and Mitigation of Microbiologically Influenced Corrosion in the Natural Gas Industry," Gas Research Institute Report GRI-92/0382 (Chicago, IL: Gas Research Institute, 1992).
26. Pope, D.H., D.M. Dziewulski, T.P. Zintel, and O.W. Siebert, *Guide to the Investigation of Microbial Corrosion in Gas Industry Facilities* (Chicago, IL: Gas Research Institute, 1988).

27. Pope, D.H., T.P. Zintel, A.K. Kuruvilla, and O.W. Siebert, "Organic Acid Corrosion of Carbon Steel: A Mechanism of Microbiologically Influenced Corrosion," CORROSION 1988, paper no. 88079 (Houston, TX: NACE International, 1988).
28. Eckert, R., *Field Guide for Investigating Internal Corrosion of Pipelines* (Houston, TX: NACE International, 2003).
29. Eckert, R.B., *Field Guide to Internal Corrosion Mitigation and Monitoring for Pipelines* (Houston, TX: NACE International, 2016).
30. Eckert, R.B., "Note on Microbiological Analyses," correspondence to author, March 29, 2017.
31. Powell, D., L. Stapley, J. Hughes, and E. Brown, "Controlling the Formation of a Non-Sulfur Based Black Powder Within a New Natural Gas Processing Plant's Waste Recovery Unit and Minimizing any Threats of a Possible Metal Dusting Corrosion Mechanism Within Process Tubing," CORROSION 2012, paper no. C2012-0001438 (Houston, TX: NACE International, 2012).
32. Zhang, P., Z. Han, C. Ye, and L. Zhu, "Problems and Mitigation Experiences with Black Powder Inside of the Long Distance Natural Gas Pipelines," CORROSION 2012, paper no. C2012-01589 (Houston, TX: NACE International, 2012).
33. J. Boivin, "ICMP, Task 1.3 Chemical and Bacterial Fluid Analysis, Strategy for Sample Collection and Analysis, Corrosion and Scale Deposits" (Kuwait Oil Company, 2001).
34. U.S. Department of Labor, Occupational Safety and Health Administration, OSHA Quickcard, "Hydrogen Sulfide (H_2S)" https://www.osha.gov/OshDoc/data_Hurricane_Facts/hydrogen_sulfide.pdf (accessed June 12, 2018).
35. U.S. Department of Labor, Occupational Safety and Health Administration, "Fact Sheet Hydrogen Sulfide (H_2S)," https://www.osha.gov/OshDoc/data_Hurricane_Facts/hydrogen_sulfide_fact.html (accessed June 12, 2018).
36. U.S. Department of Labor, Occupational Safety and Health Administration, "Safety and Health Topics, Hydrogen Sulfide, Hazards," https://www.osha.gov/SLTC/hydrogensulfide/hydrogensulfide_found.html (accessed June 12, 2018).
37. Oil and Gas, "Automatic Removal of Black Powder Extends Pipeline Life," https://www.oilandgasproductnews.com/article/19982/automated-removal-of-black-powder-extends-pipeline-life (accessed June 12, 2018).
38. Tsochatzidis, N.A., and K.E. Maroullis, "Methods Help Remove Black Powder from Gas Pipelines," *Oil and Gas Journal* (March 2007): pp. 52–58.
39. Baldwin, R.M., in discussions with (author) D.E. Powell, Private Communications, May 18, 2017.
40. Powell, D.E., D. Ma'Ruf, and I. Rahman, "Practical Considerations in Establishing Corrosion Monitoring for Upstream Oil and Gas Gathering Systems," *Materials Performance,* August 2001 (Houston, TX: NACE International, 2001).
41. Roberge, P.R., "Corrosion Monitoring," *Uhlig's Corrosion Handbook,* 3rd ed., R. Winston Revie, Editor (Hoboken, NJ: John Wiley and Sons, Inc., 2011).
42. ASTM D 513-82, "Total and Dissolved Carbon Dioxide in Water, Test Method E" (West Conshohocken, PA: ASTM International).

43. McBride, R.S., and J.D. Edwards, "Lead Acetate Test for Hydrogen Sulphide in Gas," Technologic Paper, No. 41, August 19, 1914. https://nvlpubs.nist.gov/nistpubs/nbstechnologic/nbstechnologicpaperT41.pdf (accessed June 12, 2018).
44. ASTM D 4084-07-2012, "Standard Test Method for Analysis of Hydrogen Sulfide in Gaseous Fuels (Lead Acetate Reaction Rate Method)" (West Conshohocken, PA: ASTM International, 2012).
45. U.S. Environmental Protection Agency, "Methods for Chemical Analysis of Water and Wastes, Method 376.2 (Washington, DC: US EPA, 1983).
46. NACE SP0192-2012 "Monitoring Corrosion in Oil and Gas Production with Iron Counts" (Houston, TX: NACE International, 2012).
47. NACE TM-0194-2014, "Field Monitoring of Bacterial Growth in Oil and Gas Systems" (Houston, TX: NACE International, 2014).
48. Fischer, H., "X-Ray Fluorescence Spectroscopy (XRF)–Basics," www.xrf-spectroscopy.com (Windsor, CT: Fischer Technology, Inc., 2017): accessed June 12, 2018.
49. AMPTEK, Inc., "Low Energy Fluorescence Spectra, Image," used with permission, www.amptek.com (2017).
50. "X-ray Crystallography," Available at URL https://en.wikipedia.org/wiki/X-ray_crystallography (accessed February 27, 2018).
51. U.S. Geological Society, "X-Ray Diffraction Primer," https://pubs.usgs.gov/of/2001/of01-041/htmldocs/xrpd.htm (accessed February 27, 2018).
52. Powell, D., R. Einer, J. Cheek, D. Fincher, R. Gonzales, R.H. Winters, B. Rodrigue, and B. Means, "Practical Experiences in Re-Establishing Control over SRBs and Bacteria Generated H_2S in Subsea Pipelines following Hurricane Ike," CORROSION 2010, paper no. 10064 (Houston, TX: NACE International, 2010).
53. Powell, D.E., D. Fincher, R. Gonzales, B. Melancon, R.H. Winters, B. Rodrigue, and E. Brown, "Beneficial Effects of Chemical Treatments and Maintenance Pigging Program in Returning an Offshore Pipeline to Pre-Hurricane Ike Conditions Following a Breach and the Ingress of Seawater and Sand, and the Effects of Bacteria Generated H_2S," CORROSION 2012, paper no. C2012-0001098 (Houston, TX: NACE International, 2012).
54. NACE SP-0775-2013, "Preparation, Installation, Analysis, and Interpretation of Corrosion Coupons in Oilfield Operations" (Houston, TX: NACE International, 2013).
55. Powell, D., M. Bohon, and G. Chestnut, "Solid Gel Pigs for Cleaning Production Pipelines," CORROSION/96, paper no. 96048 (Houston, TX: NACE International, 1996).
56. Powell, D.E., B. Melancon, and R.H. Winters, "Guidelines When Conducting Hydrostatic Pressure Tests and Management of Extended Dwell Times Through Dry/Wet Lay-Ups," CORROSION 2011, paper no. 11082 (Houston, TX: NACE International, 2011).
57. NACE Basic Corrosion Course (Houston, TX: NACE International, 1996).
58. Eckert, R.B., H.C. Aldrich, C.A. Edwards, and B.A. Cookingham, "Microscopic Differentiation of Internal Corrosion Initiation Mechanisms in Natural Gas

Pipeline Systems," CORROSION 2003, paper no. 03544 (Houston, TX: NACE International, 2003).
59. Rohrback Cosasco Systems, *Corrosion Monitoring Primer* (Santa Fe Springs, CA: Rohrback Cosasco Systems, 1995): p. 16.
60. Powell, D.E., D. Ma'Ruf, and I. Rahman, "Field Testing Corrosion Inhibitors in Oil and Gas Gathering Systems," *Materials Performance* 58 (2002): pp. 42–45.
61. Powell, D.E., "Internal Corrosion Monitoring Using Coupons and ER Probes," in *Oil and Gas Pipelines Integrity and Safety Handbook*, R. Winton Revie, Editor (Hoboken, NJ: John Wiley and Sons, 2015).
62. Trifilleff, O., and T.H. Wines, "Black Powder Removal from Transmission Pipelines: Diagnostics and Solutions," http://www.pall.de/pdfs/Fuels-and-Chemicals/FCBLACKPEN.pdf (accessed June 12, 2018).
63. Yamada, J., H. Kaneta, and K. Nakayama, "Analysis of Black Powder in Natural Gas Pipeline," CORROSION 2011, paper no 11088 (Houston, TX: NACE International, 2011).
64. Perry, J.H., "Characteristics of Dispersed Particles," *Chemical Engineers Handbook*, 3rd ed., Figure 113 (New York, NY: McGraw Hill Book Company, 1950).
65. Mueller, F., "Fundamentals of Gas Solids/Liquids Separation," www.muellerenvironmental.com/res/uploads/.../200-059-GMRC-2004-Separation.pdf (accessed June 12, 2018).
66. Winters, R.H., "Black Powder," NACE Tulsa Presentation (Houston, TX: NACE International, 2013).
67. Winters, R.H., and D.E. Powell, Personal Discussions and Correspondence (May 18, 2017).
68. Baroni, S., "The Effects of Cleaning for Inspection on Gas Pipeline Efficiency—A Hydraulic Modeling Perspective," Pipeline Pigging and Integrity Management Conference, February 17, 2010 (Houston, TX: Clarion, 2010).
69. Wicks, M., "Transportation of Solids at Low Concentrations in Horizontal Pipelines," in *Advances in Solid-Liquid Flow in Pipelines and its Applications*, Iraj Zandi, Editor (New York, NY: Pergamon Press, 1971): pp. 101–124.
70. Smart, J.S., in discussions with (author) D.E. Powell, Private communications (September 25, 2017).
71. Clariant Chemicals, "H_2S Scavenger," Oilfieldwiki.com. www.oilfieldwiki.com/wiki/H2S_scavengers# (accessed June 12, 2018).
72. Gilbert, P.D., J.M. Grech, R.E. Talbot, M.A. Veale, K.A. Hernandez, "Tetrakishydroxymethylphosphonium sulfate (THPS) for Dissolving Iron Sulfides Downhole and Topside–A Study of the Chemistry Influencing Dissolution," CORROSION 2002, paper no. 02030 (Houston, TX: NACE International, 2002).
73. Endean, H.J., *Oil Field Corrosion Detection and Control Handbook* (Houston, TX: Champion Technologies, Inc., 1989).
74. Anthony, J.W., R.A. Bideaux, K.W. Bladh, and M.C. Nichols, Editors, *Handbook of Mineralogy* (Chantilly, VA: Mineralogical Society of America, 1990).

75. Wikipedia, s.v. "Pyrrhotite," "Troilite," "Mackinawite," "Pyrite," "Marcasite," "Ferric Sulfide," "Smythite," and "Greigite," http://en.wikipedia.org/ (accessed July 30, 2018).
76. Fowler, T.A., P.R. Holmes, and F.K. Crundwell, "Mechanisms of Pyrite Dissolution in the Presence of *Thiobacillus ferrooxidans*," *Applied and Environmental Microbiology*, 65, 7 (1999): pp. 2987–2993.
77. "Glossary for Internal Corrosion," NACE Course, Internal Corrosion for Pipelines—Basic, Second Appendix (Houston, TX: NACE International, 2008).

A General Rules of Thumb

OSHA Permissible Exposure Limits for H_2S:

- General construction/shipyard 8 h limit — 10 ppm
- General ceiling limit — 20 ppm
- General industry peak limit — 50 ppm (up to 10 min if no other exposures during shift)
- Loss of smell — 100–150 ppm
- Staggering, collapse in 5 min — 500–700 ppm
- Death after 30–60 min — 500–700 ppm

H_2S May Generate Significant Volumes of Black Powder

- 4 ppm H_2S in 10 MMscfd can generate 3500 lb iron sulfide per year
- 4 ppm H_2S in 100 MMscfd can generate 70,000 lb (35 tons) in two years
- 4 ppm H_2S in 283,000 m^3/d can generate 1588 kg iron sulfide per year
- 4 ppm H_2S in 2,830,000 m^3/d can generate 31,750 kg iron sulfide in two years

Pipeline Cleanliness

- Clean pipeline = less than 2 lb solids per MMscf of gas (< 3.2 kg per 100,000 m^3)
- Very dirty pipeline = 300 to 400 pounds per MMscf in severe cases (480–641 kg/100,000 m^3)

Categories of Particle Size

- Course particles — 10 μm or greater
- Fine particles — less than 10 μm
- Smoke (e.g., cigarette) — particles from 0.25–1.0 μm

Types of Separators from Perry's Engineering[64]*

- Air filters — 2–50 μm
- Cloth collectors — 0.05–50 μm
- Scrubbers — 0.1–100 μm
- Centrifugal separators — 2–1,000 μm
- Gravity settling chambers — 5–10,000 μm

*Obtain most current specifications for separators from their manufacturers

Relative Surface Roughness:

Coating Status	Roughness (Imperial)	Roughness (Metric)
Internally coated (New)	< 250 μin	< 6.35 μm
Standard noncoated (New)	≤ 1800 μin	< 45.7 μm
Noncoated after 10 years (average)	~ 4,200 μin	~ 106.7 μm
Potential for corroded pipe (internal)	~ 18,000 μin	~ 457.2 μm

Corrosion Rate Classifications per NACE SP-0775[54]

	Average Corrosion Rate		Maximum Pitting Rate	
	mm/y	mpy	mm/y	mpy
Low	< 0.025	< 1.0	< 0.13	< 5.0
Moderate	0.025-0.12	1.0-4.9	0.13-0.20	5.0-7.9
High	0.13-0.25	5.0-10	0.21-0.38	8.0-15.
Severe	> 0.25	> 10	> 0.38	> 15

Pipeline Specs

- Nominally dry pipeline < 7 lbs/MMscf (< 0.112 g/m^3)
- H_2S typically less than ¼ grain H_2S per 100 ft^3, or approximately 4 ppm H_2S

Partial Pressure of CO_2 and Potentially Corrosive Conditions

- P_{CO2} > 30 psi (> 206.8 kPa) indicates corrosive conditions probable
- P_{CO2} between 7 and 30 psi (48.3 and 206.8 kPa) indicates corrosive conditions may exist
- P_{CO2} < 7 psi (< 48.3 kPa) indicates generally noncorrosive conditions

Limiting Velocities[73]

Well Stream Condition	1000 psi (6895 kPa)	5000 psi (34,475 kPa)
Wet, noncorrosive	85 ft/s (25.9 m/s)	75 ft/s 22.9 m/s)
Wet, corrosive (m/s)	50 ft/s (15.2 m/s)	40 f/s (12.2
Wet, corrosive, abrasive	30 ft/s (9.1m/s)	25 f/s (7.6 m/s)

Elements of Corrosion Mitigation Strategies and Integrity Management

- Frequent maintenance pigging
- Use of line drips or scrubbers to remove particulates
- Internal coatings
- Adequate separator and scrubber capacity at processing facilities
- Pipeline cleaning using chemical products
- Corrosion inhibitors, biocide treatments
- Corrosion Monitoring by metal coupons or electronic probes
- Inspection Programs, including RT, UT, MFL, ILI

B Types of Iron Sulfide

Adapted from a chart courtesy of the Mineralogical Society of America,[74] *with additional information sourced from Wikipedia*[75] *and R.M. Baldwin.*[2]

Mineral Name	Formula	Properties	Additional Information
Pyrrhotite 	$Fe_{(1-x)}S$ x = 0–0.2	Hardness: 3.5–4.5 Density: ≈4.84 gm/cm^3 Color: bronze-to-dark brown Weakly ferromagnetic. Mineral is usually massive or granular with no cleavage.	Pyrrhotite is a type of FeS that forms in sour environments including oilfields. Pyrrhotite is a nonstoichiometric variant of FeS, which is iron deficient, meaning less iron than sulfur in the monoclinic structure; second strongest naturally occurring magnetic mineral after magnetite, it can attract a paper clip.
Troilite 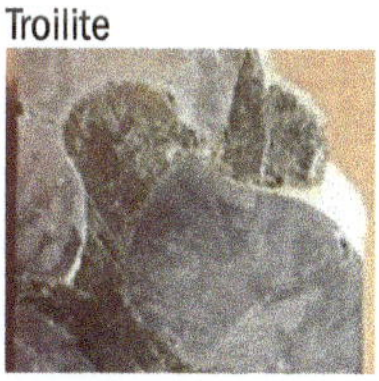 Polished meteor	FeS	Hardness: 3.5–4 Density: ≈4.61 gm/cm^3 Color: bronze-to-brown (weathered). Troilite is the iron rich, end member of the pyrrhotite group. Hexagonal crystal.	Troilite is a nonmagnetic form of iron sulfide. The hexagonal crystal does not have iron vacancies within the structure. Rare on Earth, but found in meteors originating on the Moon and Mars; common on the Moon, confirmed by the *Apollo*, *Viking*, and *Phobos* space probes, also on moons of Jupiter and in volcanic minerals on Earth.

Mackinawite 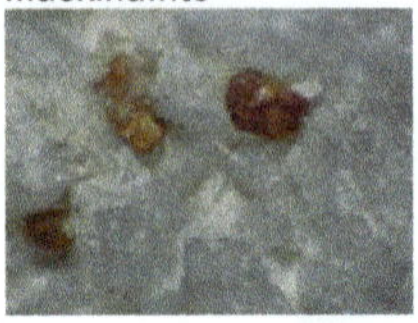	$Fe_{(1+x)}S$ x = 0-0.11	Hardness: 2.5 Density: ≈4.17 gm/cm^3 Color: bronze-to-gray white color. Tetragonal crystal. Mackinawite is a semi-stable form of FeS that forms under conditions of low H_2S activity in slightly sour environments.	Mackinawite is a tetragonal crystal, which has an excess of iron, relative to the sulfur content. In nature, usually found with nickel as $(Fe, Ni)_{1+\text{-}}S$. It can transform into greigite or pyrite, depending upon the localized oxidation/reduction environment.
Pyrite 	FeS_2	Hardness: 6-6.5 Density: ≈4.9-5.1 gm/cm^3 Pyrite is the most common form of iron sulfide, known as "Fool's gold." It has a simple cubic structure. Pyrite was formerly used in flintlock pistols as the source for sparks to ignite the gunpowder.	Pyrite is an ordered solid solution of iron and sulfur. In the natural environment, it is continuously being formed and decomposing. It breaks down into iron oxide and iron sulfate. *Acidithiobacillus* sp. bacteria can quicken degradation of pyrite.[76] Decomposition is much more rapid when in fine crystals or dust.
Marcasite 	FeS_2	Hardness: 6-6.5 Density: 4.8 gm/cm^3 Called "white iron pyrite," it has Orthorhombic crystalline structure. Marcasite is lighter and more brittle than pyrite. The hardness is such it cannot be scratched by a knife.	Marcasite has the same chemical formula as pyrite, but has an orthorhombic, rather than a cubic structure. It is more brittle than pyrite, and easily breaks up due to the unstable crystal structure. It has a pale yellow-to-white appearance. When joined in groups, the thin, flat, tabular crystals are called "cockscombs."
Ferric Sulfide	Fe_2S_3	Density: 4.3 gm/cm^3 Ferric Sulfide is a relatively unstable compound that decomposes at ambient temperatures into a yellow-greenish powder.	A solid, black powder that decays at ambient temperature (> 20 °C) into a yellow-green powder, consisting of FeS and S. Ferric Sulfide does not occur in nature.

Smythite 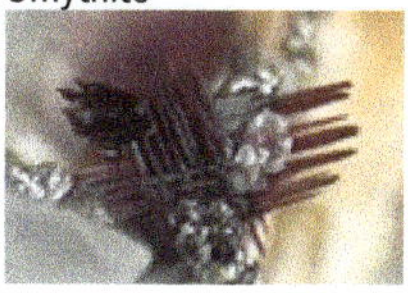	$Fe_{(3+x)}S_4$ (x = 0–0.3)	Sp. Gr.: 4.06 g/mL Hardness: 4.5 Color: greyish yellow-to-reddish black Rhombohedral structure. Strongly ferromagnetic.	One of the three iron sulfides besides FeS and FeS_2. Smythite is stable at high temperatures. It is commonly found with nickel in nature. If found within carbon steel pipelines, check integrity of stainless steel components upstream.
Greigite 	$(Fe^{2+}Fe^{3+})_2S_4$ or Fe_3S_4	Hardness: 4.5 Density: ≈4.05 gm/cm^3 Cubic crystal. Ferrimagnetic. Color: Pale pink-to-metallic blue-black when tarnished. Greigite is formed by magnetotactic bacteria and sulfate-reducing bacteria (SRB).	Greigite is a combination of Iron (II) and Iron (III) sulfide in repeating centers in a 1:2 ratio. It is the sulfur equivalent of the iron oxide magnetite (Fe_3O_4), with sulfur replacing the oxygen. It has a spinel structure. It has not been possible to form greigite by direct reactions between iron and sulfur in laboratories.
Higher Multiples			Usually forms of pyrrhotite or mackinawite.

C

Worksheet for Basic Data on Black Powder Samples

The following worksheet was developed to provide a format for an industry-wide database on black powder samples.[8] It provides the basic data fields, which could be used by operators and pipeline companies to document the collection of black powder samples.

"Black Powder" Sample and Database Information

Contact information: ______________________________

Company: ______________________________

*Address, City, State/Province, Country: ______________________________

*Pipeline name or designation: ______________________________

*Knowledgeable person: ______________________________

*Phone no.: ______________ and/or E-mail address: ______________

Sample information: ______________________________

*How long (time) between opening the pressure boundary exposing the sample to air, and it being sealed in this container? ______________ (hours, minutes)

*What gas or liquid was used to purge this sample? ______________________________

Hardware from which the sample was removed:____________________________

How long since this pipe line was cleaned or pigged?______(months)

*Average or typical hydrogen sulfide (H_2S) content of the gas in this pipeline?
________parts per million, or __________ grains per 100 cubic feet of gas.

Average or typical total sulfur content of the gas in this pipe line?
____________ grains per 100 cubic feet (ppm) of gas.

*Average or typical flow rate through this pipe? ________________ scfd (m^3/d), or specify units.

Pipeline gas temperature and pressure at this site (give units)
T______________________, P ______________

Acceptable moisture content? ___________ lb per MMscf (g/m^3) of gas. Is this often exceeded? _______

Digital photographs of the collection would be of interest.

E-mail to: ______________________________

Glossary of NACE Internal Corrosion Terms

Adapted from the NACE International Course "Internal Corrosion for Pipelines–Basic"[77]

Acid-Producing Bacteria (APB):	Bacteria that produce Organic Acids as an end product of their metabolism, which may be aerobic or anaerobic.
Aerobic:	Containing or utilizing air or oxygen.
Aerobic Bacteria:	Bacteria that require oxygen for metabolism and growth.
Alkalinity:	A measure of water's ability to neutralize acids.
Anaerobic:	Free of air or oxygen.
Anaerobic Bacteria:	Bacteria able to grow without free oxygen; oxygen is toxic to many of these organisms.
Anion:	A negatively charged ion (e.g., chloride–Cl^-, sulfate–SO_4^{2-}, etc.) that reacts with positively charged ion species (see "Cation") to form salts or other compounds, which can form scale deposits, and may promote or inhibit corrosion.
Anode:	The electrode of an electrochemical cell at which oxidation is the principal reaction and corrosion occurs.
Aqueous:	A liquid containing water.
Average Pit Density:	The average number of pits per square centimeter (cm^2) located on any side of an Electron Microscope Coupon as observed through a light microscope at a magnification of 10×. This measurement may range from 0 to 99, with a default value of 100 (the maximum detection value) indicating when 100 pits/cm^2 or greater are counted.

Batch Treatment: Sometimes known as slug treatment, performed by injecting corrosion inhibitor or biocide at one location or various selected points on a system. Generally used in conjunction with pigging operations. Normally works best where no free liquids are transported with the gas.

Bicarbonate (HCO_3^-) Ion: Is naturally occurring in some oilfield waters and can act as a buffer and prevent the acidity in the water from increasing (see "Alkalinity") and may also promote scale formation.

Biocides: An additive used to kill or control bacteria. Some biocides also have inhibiting powers in certain corrosive environments.

Calcium (Ca^{++}) Ion: Calcium is an "active metal" cation that is a common dissolved constituent of most waters that represents a percentage of the total dissolved solids and is a hardness component of water. Calcium ions can combine with sulfate or carbonate ions to form insoluble compounds, which may in turn form deposits and scales that may be protective. Excessive and uncontrolled formation of scale deposits can decrease the diameter of the pipe and create flow problems.

Carbonate (CO_3^{2-}) Ion: Naturally occurring ions in some oilfield waters that participate in the buffering of acids and in the formation of scales.

Carbon Dioxide (CO_2): A colorless odorless gas that is present at varying levels in natural gas. When dissolved in soft water, CO_2 forms carbonic acid (H_2CO_3), which is corrosive to iron and steel. When hard water is present, the CO_2 will form carbonates (CO_3^{2-}) and/or bicarbonates (HCO_3^-).

Cathode: The electrode of an electrolytic cell at which reduction is the principal reaction.

Cathodic Protection (CP): A technique to reduce the corrosion rate of a metal surface by making it the cathode of an electrochemical cell. The corrosion rate can be reduced by shifting the corrosion potential of the electrode toward a less oxidizing potential by applying an external Electromotive Force or by using a sacrificial anode.

Cation: A positively charged ion (e.g., sodium-Na^+, iron–Fe^{2+} or Fe^{3+}) that reacts with negatively charged ion species (see "Anion") to form salts or other compounds, which can form scales and deposits.

Cavitation: It is the erosion of a surface due to the sudden formation and collapse of bubbles that contain vapor, gas, or both resulting from liquids moving at high velocities. As the bubbles collapse and the surrounding liquid surfaces meet, a great deal of kinetic energy is released and may break a protective surface film on the metal and lead to the beginnings of metal loss/corrosion.

Check Reading: A reading to test an ER probe using the electronic meter's two internal references, where the difference of the two should remain constant (relative to temperature).

Chloride (Cl^-) Ion: An anion that is a very common dissolved constituent of most waters and is known to promote corrosion, particularly localized (pitting, crevice, etc.) corrosion. Chloride typically represents the most significant percentage of the total dissolved solids in many water samples because it is the chief component of most brines.

Coating: A material (usually organic) applied to a structure to separate it from the environment.

Concentration Cell: An electrolytic cell, the EMF of which is caused by a difference in concentration of some component in the electrolyte. This difference leads to the formation of discrete cathode and anode regions.

Conductivity: The ability of a substance (measured in ohm-cm) to conduct an electric charge or current due to the presence of positively or negatively charged ions. Theoretically, the higher the conductivity of an electrolyte the greater the chance for corrosion to occur.

Continuous Injection: A method of applying corrosion inhibitors or biocides by continuous injection of chemical. Can be used in either stratified or annular flows. Generally recommended when the free liquids are transported in the gas stream.

Corrosion: The chemical or electrochemical reaction between a material, usually a metal, and its environment that produces a deterioration of the material and its properties.

Corrosion Potential (E_{corr}): The potential of a corroding surface in an electrolyte relative to a reference electrode measured under open-circuit conditions. Also called: native potential, rest potential, open-circuit potential, or freely corroding potential.

Corrosion Coupon: A small, carefully weighed and measured specimen of metal that is used to determine metal loss caused by corrosion over a specified period of time. Coupons are generally made of the same metal composition as the structure in which they are installed.

Corrosion Rate: The amount of corrosion occurring in unit time. Generally expressed in mils per year (mpy) for uniform thickness changes or metal loss. A mil is 0.001 inch, or one one-thousandth of an inch. Coupon corrosion rates may help to indicate the severity of corrosion in a system and are calculated using the known values of: W = weight loss/gain (milligrams–mg), D = density of specimen material (grams per cubic centimeter–g/cm^3), A = area of coupon specimen (inches squared– in^2), and T = exposure time (days) by the following equation:

$$mpy = W \times 365 / D^* \times 16.4 \times A \times T$$

*Density of carbon steel = 7.85

Crevice (Contact) Corrosion: Localized corrosion (similar to pitting) of a metal surface at, or immediately adjacent to, an area that is shielded from full exposure to the environment resulting in an area where flow is stagnated. Shielding may result under washers, flange gaskets, and sediments.

Critical Velocity: The velocity at which erosion-corrosion begins in a pipeline.

Culture Bottles: Bottles/vials of liquid media formulated to cultivate and quantitate different groups of bacteria found in oilfield/pipeline environments.

Dry System: A pipeline transporting an acceptable dew point gas, which will not condense free water into the gas stream.

Electrical Isolation: The condition of being electrically separated from other conductors (e.g., metals).

Electrolyte: The soil or liquid adjacent to and in contact with a buried or submerged system, including the moisture and other chemicals contained therein. In the electrolyte, the ions present will migrate in an electric field.

Electromotive Force Series (EMF Series): A list of elements arranged according to their standard electrode potentials, with "noble" metals such as gold being positive and "active" metals such as zinc being negative.

End Weight: The final weight, in grams (g), of a corrosion coupon after exposing it to pipeline conditions and cleaning. This value is used with the "Start Weight" to determine weight loss (or gain). This weight difference (typically converted to milligrams, mg) is used to calculate the corrosion rate. Note that the end weight may be affected by things other than metal corrosion, including the presence of remaining scale or the removal of metal from the coupon as a result of cleaning or mechanical damage.

Energy-Dispersive Spectroscopy (EDS): When an electron beam (i.e., SEM) is used to bombard the metal surface or solid sample, x-rays are emitted with a characteristic energy. The characteristic energy peaks emitted then are used to identify individual elements on a qualitative (i.e., not quantitative) basis.

Erosion: The progressive loss of material from a solid surface due to mechanical interaction between that surface and a fluid, a multi-component fluid, or solid particles carried with the fluid. In simple erosion, corrosion makes no contribution to the metal loss. Erosion may appear as elongated pits or channels parallel to the flow direction or general metal loss at section changes, bends or elbows. Damage due to erosion generally has a smooth appearance.

Erosion Corrosion: A conjoint action, involving corrosion and erosion in the presence of a moving corrosive fluid, leading to the accelerated loss of material. In this mechanism, erosion serves to prevent protective scales or films from forming over actively corroding areas. It may occur where changes in a section of pipe cause turbulence, or at bends and elbows. Erosion-corrosion is closely related to impingement corrosion.

Electrical Resistance (ER) Probes: An electronic probe that can be used in systems where gas or liquids (including hydrocarbons) are present to determine metal loss over time by measuring the increase in the resistance of the electrode as its cross-sectional area is reduced by corrosion. The resistance of the electrode is then compared with the resistance of a standard, non-exposed electrode.

Embrittlement: The severe loss of ductility or toughness, or both, of a material, usually a metal or alloy.

Exposure Period: The number of days a coupon is exposed to system conditions. This value is used to calculate the general corrosion rate and pitting rate.

Facultative Bacteria: Microorganisms that are capable of growing with or without the presence of a specific environmental factor (e.g., facultative aerobic). When no oxygen is available, such as in natural gas systems or in the anaerobic zone of a biofilm, facultative aerobic organisms utilize sulfates or nitrates as electron acceptors in lieu of O_2.

Flow Velocity: A calculation of the speed a gas or liquid moves through a pipe of a specific side diameter.

Free Liquids: Liquids (hydrocarbon or water) that are not vaporized or entrained in the gas phase.

Galvanic Anode: A metal that preferentially corrodes and provides protection to another metal, as in a sacrificial anode. Also applies to metals that are more noble in the series when coupled in an electrolyte because of its relative position in the galvanic series.

Galvanic Corrosion: Accelerated corrosion of a metal because of an electrical contact with a more noble metal or nonmetallic conductor in a corrosive electrolyte. Also known as dissimilar metal corrosion.

Galvanic Series: A list of metals and alloys arranged according to their relative corrosion potentials in a given environment.

Gas Flow: **Stratified Flow**–Low gas velocity (liquid concentration on bottom of pipe)

Wave Mist–Intermediate gas velocity (liquid concentration partially entrained in the gas flow)

Annular Mist–High gas velocity (liquid concentration around circumference of pipe and entrained in the gas flow)

Grind out (shake out): To spin a sample of oil at high speed to determine its basic sediment and water content.

Holiday: Any discontinuity or bare spot in a coated surface.

Hydrogen Sulfide (H_2S): A colorless, poisonous, and flammable gas that has a characteristic rotten egg odor at low concentrations. H_2S is very soluble in water. When dissolved it behaves as a weak acid and usually causes pitting of carbon steel, depending on the type of film formed on the metal surface.

Hydroxide (OH^-) Ion: A dissolved constituent of water that has the ability to neutralize acids and act as an anionic inhibitor. OH^- ions also make up a component in water's calculated alkalinity.

Impingement Corrosion: A form of erosion-corrosion generally associated with the local impingement of a high-velocity, flowing fluid against a solid surface. This type of attack occurs when a liquid stream collides with a metal surface and breaks down protective films in very small areas. Impingement is very similar to erosion-corrosion; but it is usually considered to be due to more direct blasting of the surface by a fluid, which is often the result of turbulence surrounding small particles adhering to the metal surface.

Impressed Current: An electric current supplied by a device employing a power source that is external to the electrode system. An example is direct current for CP.

Inhibitors: A chemical substance or combination of substances that, when present in the proper concentration and forms in the environment, prevents or reduces corrosion. Inhibitors have solubility and dispersibility characteristics, and some are pH-activated. Depending on the product, concentration levels of 50 to 250 ppm are desired residuals in the appropriate liquid phase.

In-Line Inspection ("Smart Pigging"): The inspection of a steel pipeline using an electronic instrument or tool that travels along the interior of the pipeline in order to locate corrosion and/or material defects.

In situ: A Latin phrase meaning "in place." Any process that occurs in the field as opposed to the factory. Internal coating of pipe after installation is called in situ coating.

Intergranular Corrosion: Preferential corrosion at or adjacent to the grain boundaries of a metal or alloy, with relatively little corrosion of the grains. This occurs when there are dissimilarities in the activities between the grain boundaries and the grain that can be caused by impurities at the grain boundary, enrichment of one of the alloying elements, or depletion of one of these elements in the grain boundary areas.

Iron (Fe^{2+}/Fe^{3+}) Ion: Iron is a metallic cation that is a common dissolved constituent of most waters (generally in the oxidized ferric–Fe^{3+} form) and the chief indicator of corrosion of carbon steel. In anaerobic environments such as natural gas systems, iron may also be present as the reduced ferrous (Fe^{2+}) form. Dissolved iron: Generally determined in the field on untreated aqueous samples using colorimetric methods, it indicates all iron species in solution. Total iron: Generally determined in the laboratory on acidified samples, which solubilize any undissolved iron in the sample for detection of all iron present.

Iron Carbonate ($FeCO_3$): A corrosion product of iron and steel. $FeCO_3$ scale deposits often form when dissolved iron in water reacts with carbonates and precipitate on metal surfaces. Iron carbonate on the external portion of the pipe is an indicator that CP is reaching that area.

Iron Oxide (Fe_2O_3): A very common corrosion product of iron and steels (i.e., "rust"). Fe_2O_3 scale deposits form when dissolved iron ions in water react with air and precipitate on metal surfaces.

Iron Sulfide (FeS): A black corrosion product of iron and steel that is corrosive itself. Originates from reactions between dissolved iron and sulfide (S^{2-}) ions that may be present from hydrogen sulfide (H_2S) in the gas or produced by bacteria.

Linear Polarization Resistance (LPR) Probe: A linear polarization resistance (LPR) probe measures corrosion rates instantaneously by utilizing an electrochemical method known as linear polarization in which the potential of the probe is polarized (changed) from its inherent corrosion potential and the resulting current is measured. This resulting current can then be used to estimate the corrosion rate.

Magnesium (Mg^{2+}) Ion: Magnesium is a very reactive, divalent metal cation. It is a common dissolved constituent of most waters that represents a percentage of the total dissolved solids and is a major hardness component of water. Magnesium ions can combine with sulfate or carbonate ions to form insoluble compounds, which may in turn form deposits and scales that may be protective. Excessive and uncontrolled formation of scale deposits can decrease the diameter of the pipe and create flow problems.

Manganese (Mn^{n+}): Manganese is a metallic cation that has multiple oxidation states and is a common dissolved constituent of most waters, but usually at relatively low concentrations. Mn is also a constituent of carbon steel. Its presence within liquid samples may not only indicate whether corrosion is occurring within the system but also can be used to follow the corrosion process.

Metabolism: Chemical reactions in all living organisms that are responsible for growth and survival. In many microorganisms for example, these reactions are oxidation-reduction reactions that are accompanied by a release of energy. Because microbes may act as electron donors or electron acceptors, microorganisms attached to metal surfaces are capable of producing electrochemical reactions that can drive a corrosion cell.

Methanol (CH_3OH): Generally used as a freeze-protectant, as a carrier in corrosion inhibitors or biocides, and in dewatering pipelines following hydrotesting.

Microbiologically Influenced Corrosion (MIC): Corrosion inhibited or accelerated by the presence or activity, or both, of microorganisms, including the effects resulting from metabolic processes. Two groups of bacteria often identified in MIC of natural gas systems are: Acid-Producing Bacteria (APB) and Sulfate-Reducing Bacteria (SRB).

Milligrams Per Liter (mg/L): A weight-volume relationship that is a commonly used unit of measurement for expressing the dissolved constituents detected in water samples. Milligrams per liter (mg/L) is sometimes incorrectly used interchangeably with parts per million (ppm).

To convert mg/L to ppm, the following equation can be used:

$$(\text{mg/L}) / (\text{specific gravity}) = \text{ppm}$$

Mixed Potential: The potential of a specimen (or specimens in a galvanic couple) when two or more electrochemical reactions are occurring simultaneously (e.g., corrosion of iron as the anodic reaction and reduction of oxygen as the cathodic reaction).

Mole Percent (mol. %): The percentage of a compound or moles in a sample. To convert partial pressure to mol. %, use the following equation:

$$\text{Mole Percent (\%)} = [\text{partial pressure (psia)} \times 100\%] / [\text{pressure (psig)} + \text{atmospheric pressure (psi)}]$$

For example, if the CO_2 partial pressure in natural gas was 5 psia and the pipeline operating pressure was 1,000 psig, the CO_2 mol. % would be calculated as follows:

$$\text{Mole Percent } (CO_2) = MP_{CO2} =$$

$$MP_{CO2} = [(5) \times (100)] / (1{,}000 + 1 \text{ atmosphere*})$$
$$MP_{CO2} = (500) / (1{,}014.7) = 0.49 \text{ mol. \%}$$

*1 atmosphere (sea level) = 14.7 psi

Obligate Anaerobes: Anaerobic bacteria that cannot grow in the presence of free oxygen or for which oxygen is toxic.

Organic Acids: Weak acids that are the end product of metabolism by a variety of microorganisms, which are corrosive to carbon steel and other metals. These acids are correctly classified as carboxylic acids (contain a carboxyl group–COOH) and are also known as short-chain fatty acids.

Oxidation: Loss of electrons by a constituent of an electrochemical reaction.

Oxygen (O_2): A colorless, odorless gas that, when dissolved in water, can serve as a cathodic reactant promoting corrosion. Although well known for its role in the corrosion of iron and steel when available (i.e., certain external environments), oxygen generally is not a factor in most internal corrosion of natural gas systems because it is not usually present in produced gas and its ingress is minimized.

pH: The negative logarithm of the hydrogen-ion concentration in a substance. A pH of 7.0 is neutral. A pH lower than 7.0 is acidic while a pH greater than 7.0 is alkaline.

Partial Pressure: The pressure, in pounds per square inch absolute (psia), of a component gas in a pipeline or vessel. The partial pressure of a gas represents the pressure that a gas would exert on a system if it were the only gas present. The partial pressure of a gas can be computed using the following equation (assuming ideal gas):

$$\text{partial pressure} = [\text{fractional mole \%} \times \text{total pressure}]$$

where:

$$\text{fractional mol. \%} = [(\text{mol. \%}) / (100)]$$

$$\text{total pressure} = [\text{pressure} + \text{atmospheric pressure}]$$

Total Pressure is abbreviated psia.
Measured Pressure in pipeline or vessel is abbreviated psi.
Atmospheric pressure (atm) at sea level is approximately 14.7 psi.

For example, if a gas sample has 0.5 mol. % of CO_2 and the pipeline operating pressure was 1,000 psig, the CO_2 partial pressure would be calculated as follows:

$$\text{Partial Pressure } (CO_2) = PP_{CO2}$$
$$PP_{CO2} = [(0.5) / (100)]\ (1{,}000 + 1 \text{ atm}) =$$
$$PP_{CO2} = (0.005)\ (1{,}014.7) = 5.07 \text{ psia}$$

Parts Per Million (ppm): One of the commonly used units for reporting water analysis data. The part per million (ppm) is a measure of proportion by weight, equivalent to a unit weight of dissolved substance per million unit weights of solution. The unit used in water analysis is the milligram (mg), so 1 ppm is equivalent to 1 mg of solute per 1,000 grams of solution.

Pitting: Corrosion of a metal surface, confined to a point or small area, that takes the form of cavities. The metal undergoing corrosion suffers localized metal loss, rather than over the entire surface. In general, a pit may be described as a cavity or hole where the surface diameter is about the same size or less than its depth.

Pitting Rate: The rate of penetration calculated using the maximum pit depth measured on a coupon. This penetration is extrapolated over a 1-year period and expressed in mils per year (mpy). The following equation is used to find the pitting rate:

$$\text{mpy} = \frac{\text{max pit depth } (\mu\text{m}) \times 0.03937 \text{ (mil/}\mu\text{m)} \times 365 \text{ (days/year)}}{\text{exposure period (days)}}$$

Planktonic Bacteria: Free-floating or free-swimming bacteria in bulk fluids.

Potassium (K^+) Ion: A monovalent metallic cation that is a common dissolved constituent of most waters and formation brines.

Pressure: The force exerted or applied over a surface, expressed in pounds per square inch gauge (psig). Pressure affects the rates of most chemical reactions, including corrosion reactions. The primary importance of pressure is its effect on dissolved gases; i.e., more gas goes into solution as the pressure is increased. This may in turn increase the corrosiveness of the solution.

Reduction: The gain of electrons by a constituent of an electrochemical reaction.

Scaling Tendency: Describes a water's potential for scale formation (see "Stability Index"). The tendency for scaling in a system is influenced by temperature, pressure, concentration of different ions, and amount of dissolved gases. A strongly positive stability index suggests a potential for scaling, whereas a very negative stability index predicts no scale precipitation and that there is a potential for corrosion. Values in between are generally considered inconclusive with regard to scaling or corrosion.

Sessile Bacteria: Bacteria attached to a surface.

Sodium (Na^+) Ion: A monovalent metallic cation that is a common dissolved constituent of water, brines, and scales. Sodium commonly forms salts with anions such as chloride. The addition of salts to water increases its conductivity and possibly its corrosiveness.

Specific Gravity: The ratio of the weight of a given volume of a substance at a given temperature (60 °F [15.6 °C] is often used) to the weight of a standard substance at the same temperature. Specific gravity of a solution can be used as an indication of the amount of salts dissolved in water. As the specific gravity increases, the density, and therefore the amount of dissolved salts, also increases.

Stability Index: An empirical expression that indicates the scaling and corrosion tendencies of water samples. The stability index provides a somewhat quantitative value for whether calcium carbonate scale will be deposited within a pipeline, as well as potential seriousness of any scaling or corrosion that might occur. In general, more positive numbers indicate a greater likelihood of scale formation and more negative numbers indicate an increased possibility for corrosion. The stability index is accurate when calculated using field measured pH and alkalinity.

Start Weight: This is the initial weight, in grams (g), of a corrosion coupon before exposing it to pipeline conditions (also known as "Begin Weight"). This value is used with the "End Weight" to determine weight loss (or gain). This weight difference (typically converted to milligrams, mg) is used to calculate the corrosion rate.

Sulfate (SO_4^{2-}): A non-metallic, anionic divalent compound that is a common dissolved constituent of most waters. SO_4^{2-} can combine with calcium or magnesium ions to form insoluble compounds, which may in turn form deposits that may serve as a protective scale. However, if the formations of these deposits are excessive and uncontrolled, their build-up can decrease the diameter of the pipe and create flow problems. Sulfate is also important because it is reduced by sulfate-reducing bacteria (SRB) to sulfide (highly corrosive) and oxygen.

Sulfate Reducing Bacteria (SRB): A group of anaerobic bacteria that reduce sulfate to sulfide (see "Microbiologically Influenced Corrosion"). Considered by many to be the chief causative agent of MIC.

Sulfur: An element that has multiple possible forms. The relevant forms that may be present in a natural gas system include H_2S, elemental sulfur, dissolved sulfide, and oxyanions of sulfur (e.g., sulfate–SO_4^{2-}).

Temperature: Temperature, in degrees Fahrenheit (°F), or degrees Celsius (°C) is a measure of the heat present in a substance or its environment. Because temperature changes rapidly, it must be measured at the time of sample collection. All chemical reactions are affected by temperature either directly or as a result of its effect on reaction rates, solubility, phase change, etc. Some of these effects are complex; however, in general, an increase in temperature increases the corrosion rate. A rough "rule of thumb" suggests that the reaction rate doubles for every 10 °C rise in temperature. To convert degrees Celsius into Fahrenheit, the formula is as follows:

$$°F = [°C \times 1.8] + 32$$

Total Alkalinity: Alkalinity is a measure of a liquid's ability to neutralize acids. Total alkalinity is the sum effect of all the major sources of alkalinity, such as carbonate (CO_3^{2-}), bicarbonate (HCO_3^-), and hydroxide (OH^-) alkalinity. Total alkalinity measurements should be performed in the field immediately following sample collection because the value is pH-dependent and capable of changing over time.

Total Dissolved Solids (TDS): The sum of all dissolved ions, both cations and anions, measured in the analysis of an aqueous liquid. This value can be used to predict a water's conductivity and whether a water is composed chiefly of brine or condensed water. This value can also be used as a check to verify the completeness of an analysis with the sum of each of the individual constituents of the sample being approximately equal to the total dissolved solids value.

Under-Deposit Corrosion: Localized corrosion under or around a deposit or collection of material on a metal surface. The deposit(s) may be due to corrosion product accumulation, precipitation of solids from the water, or microbiological activity.

Uniform Corrosion: Corrosion that proceeds at about the same rate over a large area of a metal surface.

Upset Conditions: Abnormal operating conditions. With respect to internal corrosion, the concern is when corrosive constituents are temporarily introduced to the gas stream.

Water Dewpoint: The point at which gas is saturated with water vapor and begins to condense liquid water. The water content in lb/MMscf determines the temperature and pressure at which liquid condenses.

Wet System: A pipeline transporting gas saturated with water.

Perspective on Microbiological Analysis from Rick Eckert

There are a variety of techniques available for microbiological analysis. In all cases, a sample is collected and then the analysis is performed, making these techniques both offline and indirect. Microbiological tests provide a variety of different information about a sample, depending on the method. Some methods, e.g., culture and quantitative polymerase chain reaction (qPCR,) are used to enumerate specific microorganisms or functional types of microorganisms. Other methods, e.g., next generation sequencing or denaturing gradient gel electrophoresis (DGGE), look more at the relative differences in microbial population distribution rather than enumerating specific microorganisms. Different still, some methods attempt to measure microbiological activity, using adenosine triphosphate (ATP), hydrogenase, or sulfide production in the case of sulfate reducing bacteria (SRB) and sulfate reducing archaea (SRA).

Sample handling for microbiological analysis is extremely important to avoid contamination and degradation that will affect the validity of the results. Specific test methods may require particular sample handling techniques; however, in general it is important to use sterile tools and materials when handling the samples and to keep the samples cool during handling and shipping.

Culture techniques are detailed in NACE Standard TM0194 "Field Monitoring of Bacterial Growth in Oil and Gas Systems."

A number of other techniques, including molecular microbiological methods (MMM), and their application for monitoring and evaluating internal and external corrosion are described in NACE Standard TM0212 "Detection, Testing, and Evaluation of Microbiologically Influenced Corrosion on Internal Surfaces of Pipelines" and NACE Standard TM0106 "Detection, Testing, and Evaluation of Microbiologically Influenced Corrosion (MIC) on External Surfaces of Buried Pipelines."

All microbiological test methods have advantages and limitations. Those seeking specific details on microbiological test methods and their application should refer to the aforementioned NACE standards.

Index

C

J

L

M

R

S

T

U

V

W

X

www.ingramcontent.com/pod-product-compliance
Ingram Content Group UK Ltd.
Pitfield, Milton Keynes, MK11 3LW, UK
UKHW020719041026
15590UKWH00002B/770

9 781575 903835